AF773121

MRE

Materials Research and Engineering
Edited by B. Ilschner
Volume 2

K.A. Padmanabhan · G.J. Davies

Superplasticity

Mechanical and Structural Aspects, Environmental Effects, Fundamentals and Applications

With 86 Figures

Springer-Verlag Berlin Heidelberg New York 1980

KUPPUSWAMY ANANTHA PADMANABHAN

Professor of Metallurgy,
Indian Institute of Technology, Madras (India)
(formerly Banaras Hindu University, Varanasi, India)

GRAEME JOHN DAVIES

Professor of Metallurgy,
University of Sheffield, Sheffield (U.K.)

Dr. rer. nat. BERNHARD ILSCHNER

o. Professor, Institut für Werkstoffwissenschaften der Universität Erlangen-Nürnberg

Library of Congress Cataloging in Publication Data.
Padmanabhan, Kuppuswamy Anantha, 1945– Superplasticity.
(Materials research and engineering ; v. 2)
Bibliography: p. Includes index.
1. Superplasticity. I. Davies, G. J., joint author.
II. Title. III. Series: Materials research and engineering ; v. 2.
TA418.14.P32 620.1'1233 80-13292

ISBN-13:978-3-642-81458-7 e-ISBN-13:978-3-642-81456-3
DOI: 10.1007/978-3-642-81456-3

This work is subject to copyright. All rights are reserved, whether the whole or part of the material is concerned, specifically those of translation, reprinting, re-use of illustrations, broadcasting, reproduction by photocopying machine or similar means, and storage in data banks. Under § 54 of the German Copyright Law where copies are made for other than private use, a fee is payable to the publisher, the amount of the fee to be determined by agreement with the publisher.

© Springer-Verlag Berlin, Heidelberg 1980
Softcover reprint of the hardcover 1st edition 1980

The use of registered names, trademarks, etc. in this publication does not imply, even in the absence of a specific statement. that such names are exempt from the relevant protective laws and regulations and therefore free for general use.

2362/3020 – 543210

Editor's Preface

This is the second book in a new series - "Materials Research and Engineering" - devoted to the science and technology of materials. "Materials Research and Engineering" evolves from a previous series on "Reine und Allgemeine Metallkunde", which was edited by Werner Köster until his eightieth birthday in 1976.

Although the present series is an outgrowth of the earlier one, it should not and cannot be regarded as a continuation. There had to be a shift of scope - and a change in presentation as well. Metallurgy is no longer an isolated art and science. Rather, it is linked by its scientific basis and technological implications to non-metallic and composite materials, as well as to processes for production, refining, shaping, surface treatment, and application. Thus, the new series, "Materials Research and Engineering", will present up-to-date information on scientific and technological progress, as well as on issues of general relevance within the engineering field and industrial society.

Following the general position analysis of materials in the present world as given in volume 1, now volume 2 focuses on a special topic: It provides a thorough treatment of theoretical, experimental, and applied aspects of superplasticity.

Superplastic deformation of metals and ceramics is a phenomenon as attractive from a fundamental as from an engineering point of view. The unusually high ductility suggets an underlying micromechanical mechanism basically different from normal strain hardening and creep; careful experiments to clarify the contributions from dislocation glide, grain boundary shear, and diffusonal flow, respectively, are necessary.

On the other hand, the exceptional phenomenological features of superplasticity call for practical evaluation, in particular application for metal forming operations. These technological aspects are dealt with in the present volume as well as the theoretical basis. Thus, the book by Padmanabhan and Davies promises to present up-to-date, precise, comprehensive and interesting information to both scientists and engineers.

Erlangen, Germany, January 1980 Bernhard Ilschner

Author's Preface

Superplasticity in metals and alloys is a phenomenon which by its very uniqueness has attracted the attention of scientists and technologists over a considerable number of years. The earliest observations were made in the 1920's but little development followed from them at that stage. There was a resurgence of interest in the mid 1940's in Russia but the extensive research activity of recent times was stimulated by the published researches of W.A.Backofen and his co-workers at M.I.T. beginning about 1964. The subject is now one of interest to materials scientists, metallurgists, physicists and engineers.

There has been a great deal of work devoted to both the fundamental and the technological aspects of superplasticity. The understanding of the basic mechanisms of the deformation processes involved has been increased by theoretical and experimental efforts. On the other hand, the application of experimental data to real problems has not been neglected. Alloys of considerable commercial interest have been developed and there are a number of practical forming operations currently in use which utilise superplastic deformation.

In recent years the subject has been reviewed in part in technical journals and an impressive number of research papers have been published. Superplasticity is now well-established as a research and development subject and the time seems ripe for a comprehensive discussion of the subject.

This is the objective of this monograph. It considers the theoretical and experimental aspects of superplasticity, together with areas of potential or actual application. It is intended for workers in industry and in academic institutions. It will have succeeded in its aims if research and development is stimulated or if a proper understanding of the subject is given to those who wish such an understanding without a direct involvement.

MADRAS, INDIA, January 1980 K.A. PADMANABHAN

SHEFFIELD, U.K., January 1980 G.J. DAVIES

Acknowledgements

The authors thank Professor B. Ilschner, Editor of this Springer Series on Materials Research and Engineering, for his kind invitation to write this monograph. The co-operation of Mr. M. Hofmann of Springer Verlag at every stage is gratefully acknowledged.

Grateful acknowledgement is also made to the following organizations/societies and publishers for permission to reproduce illustrations: Academic Press, American Institute of Mining, Metallurgical and Petroleum Engineers, American Institute of Physics, American Society for Metals, Chapman and Hall Limited, Dr. Riederer Verlag GmbH, Elsevier Sequoia, Illinois Institute of Technology, Pergamon Press, Pratt and Whitney Aircraft Limited, Rockwell International Limited, The Metallurgical Society of A.I.M.E., The Metals Society, The Royal Society, T.I. Superform Limited and University of California Press.

Professor G.W. Greenwood, Dr. P.M. Hazzledine and Dr. P.J. Wray have also kindly given their consent for the inclusion of illustrations from their articles.

One of the authors (KAP) wishes to thank in addition the following people: Professor T.R. Anantharaman whose gentle care and constant encouragement made book-writing a pleasure; Mrs. Gita Padmanabhan but for whose assistance and understanding at every stage this monograph could not have been completed; Professor S. Ranganathan for valuable discussions on the structure of high-angle grain boundaries; Professor P. Rama Rao, who as the senior-most member of the mechanical metallurgy group at Banaras Hindu University, was generous with his help throughout the author's stay at that University.

Thanks are due to Dr. L.R. Vaidyanath, Director, Indian Copper Information Centre, for providing office facilities for the production of the manuscript in a form suitable for photo-offset printing. Financial support for acquiring secretarial assistance was given by the Aeronautics Research & Development Board, Ministry of Defence, Government of India.

" The trillionth part has not been said;
and all that has been said but multiplies
the avenues to what remains to be said. "

Herman Melville

Contents

1. Historical Introduction

Superplasticity is the deformation process that produces essentially neck-free elongations of many hundreds of percent in metallic materials deformed in tension. High ductilities are also encountered in 'superplastic' alloys during torsion, compression and indentation hardness testing. Further, these materials are characterised by very low strengths.

Superplasticity can be induced both in materials possessing a stable, ultra-fine grain size at the temperature of deformation ($\gtrsim 0.4\ T_M$, where T_M is the absolute melting point) and in those subjected to special environmental conditions, e.g. thermal cycling through a phase change. These two categories are best described as "structural" and "environmental" superplasticity, respectively (1).

Prior to 1945 the deformation behaviour of the class of alloys that possessed "viscous properties between those of glass and pitch (2)" was known under different names. The term superplasticity became common after 1945 when the Russian word СВЕРХПЛАСТИЧНОСТЬ (pronounced as sverkhplastichnost), introduced by Bochvar and Sviderskaya(3), got so translated. The comprehensive review of Underwood(4), published in 1962, firmly established this description.

1.1 First Observations of Structural Superplasticity

Rosenhain, Haughton and Bingham(2) appear to be the first to have chanced upon a form of structural superplasticity in 1920, while experimenting among others with a near-ternary eutectic (5) of zinc, aluminium and copper. Elongations only of the order of 60% were recorded but the flow closely resembled that of the viscous material pitch. The behaviour, however, was interpreted in terms of the then popular 'amorphous cement' theory of inter-crystalline cohesion wherein it was suggested that rolling partially converted the metal into an amorphous condition and the pitch-like response was the result of flow in these amorphous regions. Commenting on this work Jeffries and Archer(6), on the other hand, attributed the effect to the presence of a very fine grain size, a condition now known to be necessary for the occurrence of structural superplasticity (see Chapter 3).

Further studies on the viscous deformation of eutectic alloys by Hargreaves (7,8) and Hargreaves and Hills (9,10) led to the slightly different intercystalline "transition" zone theory where also the flow was supposed to be confined to the intercrystalline transition regions. At about the same time Jenkins (11) established up to 400 per cent tensile elongation in cadmium-zinc and tin-lead solders deformed at approximately 0.8 of the absolute eutectic temperature. Although these were essentially creep tests, the importance of a fine grain size was appreciated. Significantly, Jenkins also noted that flow in a neck resulting from local deformation often ceased and further deformation occurred only by the formation of additional necks in other places of the tensile specimen. Thus even at this early stage stabilisation of necking was recognised as a necessary requirement for ensuring large elongations (see Chapter 2).

This series of investigations culminated in the systematic study of Pearson (12) of tin-lead and bismuth-tin eutectic alloys, under both constant load and constant stress conditions. In the first set of experiments practically neck-free elongations of up to 650 percent were recorded. Using a mechanism for load reduction at regular intervals to offset the decrease in area of cross-section accompanying tensile deformation, approximately constant stress was maintained in the second set of tests. The specimen that experienced the maximum elongation of 1950 percent in this case was photographed and this has since been included in many previous reviews of superplasticity. In view of its historic importance it is presented here as Fig.1.1. On the basis of optical microscopy of the specimen surface it was also concluded that the grain size and shape did not change during elongation and that grain-boundary offsets occurred (Fig.1.2). This led Pearson to the conclusion that flow at the grain boundaries was the important mechanism of deformation. Further, the resemblance of the non-Newtonian flow behaviour to that of "colloidal solutions" and "coarser disperse systems such as particles of microscopic size" was recognised. This statement clearly anticipates the much later rheological theories of superplasticity (see Section 4.2.1).

Another suggestion of this period which anticipated much later postulates was that of Fell (13) according to which the above anomalous behaviour was the result of both viscous <u>and</u> plastic flow.

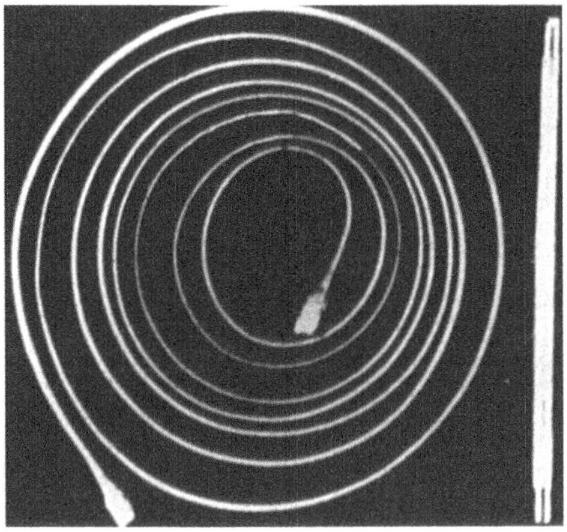

Figure 1.1 Tensile test piece of extruded bismuth-tin eutectic alloy. Elongation 1950 percent (Pearson (12))

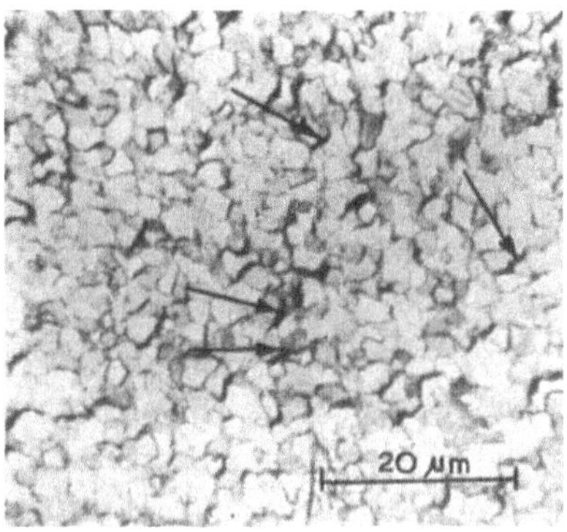

Figure 1.2 The structure of an extruded tin-lead eutectic alloy after a superplastic strain of ~ 50 percent showing grain boundary offsets (arrowed) (Pearson (12))

The papers of Pearson and Fell were the subject of much discussion (14) in which Cook (15) showed that the data of Pearson could be represented by an equation of the type

$$S = S_o + C\ V^{1/2} \qquad 1.1$$

where S is the flow stress, S_o and C are constants and V is the flow velocity (now identified with the strain rate, $\dot{\epsilon}$). Thus this is an early example of an empirical equation involving a strain-rate sensitivity index and a 'back or threshold stress' (see Sections 2.2, 2.3 and 3.1). The discussion also contains the suggestion (16) that change-in-strain-rate tests could be used to study the viscous behaviour of eutectic alloys - a technique that is presently popular in the study of superplasticity (Section 2.7.2).

Notwithstanding this advanced level of understanding of the viscous behaviour of eutectic alloys there was a gap of more than ten years before superplasticity was "rediscovered" by Bochvar and his Russian co-workers (see Underwood (4)) and a gap of thirty years before widespread interest was rekindled in the western world. It is conceivable that the impact of the then new concept of dislocations perhaps had an adverse effect on all work which depended for explanation on amorphous layers and viscous effects.

1.2 Early Observations of Environmental Superplasticity

It is now well-known that unusually high ductility can be produced by subjecting a metallic material to a small stress and then inducing a phase or allotropic transformation. For maximum effects the changes in the environment must be repeated, for example, by thermal cycling through a range which includes the transformation temperature. The earliest results in this regard were reported in 1924 by Sauveur (17, 18) using torsion tests on iron bars with an imposed longitudinal temperature gradient. It was observed that the shear deformation was localised at the site of the ferrite-austenite transformation.

Subsequently, however, attention in the United States turned towards the evaluation of hot ductility of steels and the detailed investigations reported in 1946 (19,20), although providing evidence for the presence of very high shear strains, do not seem to be concerned with processes now associated with environmental superplasticity.

During the period 1932-37 a series of papers (21-25) appeared from Germany in which enhanced ductility in steels was established when the $\alpha - \gamma$ transformation occurred under tension. A marked decrease in strength was also seen. The breaking of bonds during transformation and their re-establishment as part of the new phase was considered the important physical process. The applied stress could thus facilitate rearrangements which gave rise to deformation work. The picture was considered relevant even when the bond disruption was part of a martensitic shear transformation by postulating that out of the many possible modes of shear available only those which led to the largest elongation in the direction of the external stress were favoured. In these and a later work (26) no emphasis, however, was placed on producing large elongations.

No further results were reported till 1959 when de Jong and Rathenau (27) and Lozinsky and Simeonova (28) independently published their findings. Both these papers dealt primarily with phase transformation softening but they revived interest in transformational plasticity and this led directly to the environmental superplasticity studies of Oelschlaegel and Weiss (29). In view of the above the contemporary review of environmental superplasticity, presented in Chapter 5, starts from late 1950's. At that stage environmental superplasticity caused by two other means, viz. thermal cycling of a material with anisotropic coefficients of expansion and neutron irradiation, will also be discussed.

1.3 The Revival of Interest in Structural Superplasticity

Structural superplasticity in Zn-Al alloys was first reported in 1945 by Bochvar and Sviderskaya (3). (Here only those Russian references readily available in translation are cited. Other references may be found in Underwood (4) and in the introduction to the classical paper by Backofen, Turner and Avery (42) discussed later in this section.) In particular, the low strength and high ductility of the heat-treated "granular" zinc -20wt.% aluminium alloy were commented upon. Subsequent studies (30,31) reported elongations of several hundred per-cent and confirmed the importance of composition. A solution-precipitation theory due to Bochvar (31,32) attributed superplasticity to the directional movement of atoms in the interfaces(brought about by precipitation at some interphase boundaries and dissolution at the others, arising from local temperature fluctuations). This theory was also considered applicable to creep in polyphase systems and was the subject of some contemporary controversy. It was only by 1967, when the class of alloys which exhibited superplasticity had enlarged, that the theory was shown to be inadequate (33).

After the initial work of Bochvar there was a lull in activity until the appearance of the many papers by Presnyakov and his co-workers. A number of new superplastic systems like the eutectics of aluminium-copper (34,35), aluminium-silicon (36), magnesium-copper and tin-lead (37) were identified and knowledge of superplasticity in zinc-aluminium alloys was extended (37-39). Fig.1.3 shows the high ductilities that can be achieved in the Zn-Al system and also emphasizes the

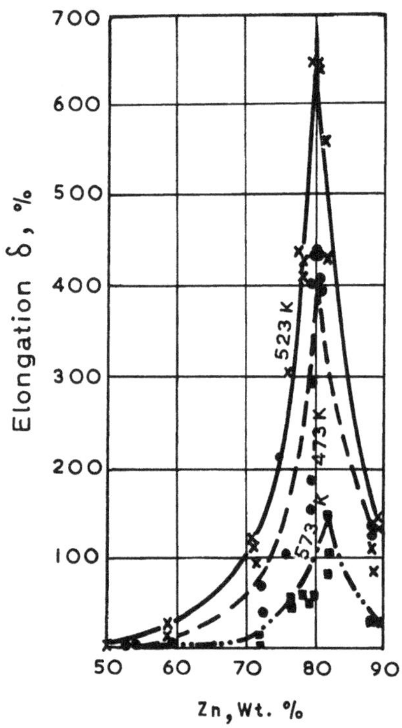

Figure 1.3 The variation of plastic elongation with test temperature for zinc-aluminium alloys of various zinc contents (Presnyakov and Chervyakova (39))

importance of the Zn-20 wt.% Al near-eutectoid alloy. The descriptions of the mechanisms of deformation, however, were vague (38). Nevertheless, these appear to be associated with a breakdown of a metastable state with superplasticity resulting from diffusion and the weakening of interatomic bonds during reconstruction of the crystalline lattice. Later papers (40,41) did not serve to clarify the position.

This then was the state-of-the art when Underwood(4) presented his review. This paper aroused considerable interest and led to the astonishing expansion of activity in researches on superplasticity. This expansion was heralded in 1964 by the paper of Backofen, Turner and Avery (42) which dealt with superplasticity in a zinc-aluminium near-eutectoid alloy. The paper also provided a phenomenological basis for superplastic flow in terms of the now well-known relationship

$$\sigma = K\dot{\epsilon}^{m} \qquad 1.2$$

where σ is the applied stress, $\dot{\epsilon}$ is the strain rate and K and m are constants dependent on test parameters like temperature and grain size. m is called the strain-rate sensitivity index and has considerable significance in determining the stability of flow. The modern view of the mechanics of superplastic deformation in which m plays a central role is given in Chapter 2. That chapter also assesses the test procedures adopted for evaluating the extent of development of structural superplasticity. The experimental and theoretical studies of structural superplasticity since 1964, on the other hand, are reviewed in Chapters 3 and 4, respectively.

As mentioned earlier both the experimental and theoretical aspects of environmental superplasticity are discussed in Chapter 5.

Moreover, in order to sustain interest in any field of research activity it is essential that fundamental researches lead to practical applications. Fortunately in case of superplasticity this has already happened and this aspect is.considered in detail in Chapter 6.

2. The Mechanics of Superplastic Deformation and the Assessment of Superplastic Behaviour

The significant suggestion of Backofen, Turner and Avery(42) that the basis of superplastic flow is to be found in the strain-rate sensitivity of flow stress is now known to apply equally to structurally as well as environmentally superplastic materials. In this Chapter the nature of the tensile test is first examined briefly. This is followed by a discussion of plastic stability and the rationalization of the phenomenology of superplastic flow in terms of a number of empirical and semi-empirical equations that have been proposed to analyse this mode of deformation. Subsequently the various methods of evaluating strain-rate sensitivity (SRS) in materials being deformed in tension will be described and also compared with other available procedures of determining this flow characteristic. The chapter will conclude with an account of some other specialised techniques, e.g. the examination of grain boundary sliding, of importance in assessing the extent of development of superplastic behaviour.

Most of the discussion will be concerned with structurally superplastic alloys which are of considerable interest. Some reference will also be made to environmentally superplastic materials, although the major discussion in this regard will be deferred until Chapter 5.

2.1 The Tensile Test

The tensile test wherein a specimen of standard dimensions is subjected to an axial (tensile) load is widely used for assessing the mechanical behaviour of materials. In the most general case the elongation of the specimen under load comprises three contributions, viz. (i) an instantaneous elastic displacement which is recovered on removal of the load, (ii) a time-dependent anelastic displacement, also recoverable on load removal and (iii) an irreversible plastic displacement which may contain both time dependent and time-independent contributions.

In general the factors that control the recoverable components (i) and (ii) are fairly well understood. The situation is, however, rather different for the plastic component.

For the plastic region the various stresses and strains can be defined as follows(43):

$$\text{nominal or engineering stress, } \sigma_n = \frac{\text{load, } P}{\text{original cross-sectional area, } A_o}$$

$$\text{nominal or engineering strain, } \epsilon_n = \frac{\text{change in length, } (l - l_o)}{\text{original length, } l_o}$$

$$\text{true stress, } \sigma_t = \frac{\text{load, } P}{\text{instantaneous cross-sectional area, } A}$$

$$\text{true strain, } \epsilon_t = \int_{l_o}^{l} \frac{dl}{l} = \ln (l/l_o)$$

It can also easily be shown that

$$\sigma_t = \sigma_n (1 + \epsilon_n) \qquad .. \quad 2.1$$

and $$\epsilon_t = \ln (1 + \epsilon_n) \qquad .. \quad 2.2$$

The nominal and the true strain rates, on the other hand, should be evaluated as the time derivatives of ϵ_n and ϵ_t, respectively.

For plastic deformation of practical interest the elastic and anelastic components can usually be neglected. This is almost always the case for superplastic materials.

Investigations relevant to the plastic range have been carried out to generate empirical, semi-empirical or theoretical equations correlating stress, strain and strain rate, and wherever possible temperature and structural variables like grain size. These equations are collectively known as mechanical equations of state (44) (or constitutive relations). Typical studies of this kind are those of Zener and Hollomon (45) and Hollomon (46) which also contain numerous references to similar enquiries.

2.2 Constitutive Relations for Superplastic Materials

In the original form of the mechanical equation of state the plastic flow stress was considered an instantaneous function of strain, strain rate and temperature T. That is

$$\sigma_t = f\ (\epsilon_t,\ \dot{\epsilon}_t,\ T) \qquad .. \quad 2.3$$

At a relatively early stage itself experiments revealed (47) that such a function did not exist if changes in temperature were allowed. It can, however, be argued (48) that for isothermal straining a relation of the form

$$\sigma_t = f\ (\epsilon_t,\ \dot{\epsilon}_t) \qquad .. \quad 2.4$$

is feasible. This is equivalent to a reduced form of the constitutive relation envisaged by Appleby (49) who showed, on the basis of physical requirements outlined by Wray and Richmond (50), that such a relation must contain at least four variables.

The first of the two general explicit forms of equation 2.4 is (51-54)

$$\sigma_t = K'\ \epsilon_t^N\ \dot{\epsilon}_t^m \qquad .. \quad 2.5$$

where K'and N the strain hardening exponent are constants. The second is (55,56)

$$\sigma_t - \sigma_y = K_1\ \epsilon_t^N\ \dot{\epsilon}_t^m \qquad .. \quad 2.6$$

where σ_y is a yield stress (or alternatively a 'back' or 'threshold' stress - see Chapter 3) and K_1 is another constant. In both the equations N and m are dependent on the test and material parameters of strain, strain rate, temperature and grain size.

For deformation at temperatures below about 0.4 T_M (T_M is the absolute melting temperature) strain hardening effects dominate. Then $m \simeq o$ and N usually lies in the range 0.1 to 0.3 (57). Then equation 2.5 reduces to

$$\sigma_t = K_2\ \epsilon_t^N \qquad .. \quad 2.7$$

For a linear Newtonian-viscous material $\sigma_y = 0$, N = 0 and m = 1 (58,59) and both equations 2.5 and 2.6 become

$$\sigma_t = K_3\ \dot{\epsilon}_t \qquad .. \quad 2.8$$

For a Bingham solid (55,58), on the other hand, N = 0 and m = 1 and equation 2.6 is written as

$$\sigma_t - \sigma_y = K_4 \dot{\epsilon}_t \qquad .. \quad 2.9$$

For superplastic deformation N is very small or zero (see Section 2.3), frequently σ_y can be neglected (60) and m is non-integral and lies in the range 0.3 to 0.9 (section 2.6). Then the empirical equation (Section 1.3)

$$\sigma_t = K \dot{\epsilon}_t^{\,m} \qquad .. \quad 2.10$$

describes the nature of the stress-strain rate relation adequately. (From a physical point of view, however, the alternative form $\dot{\epsilon}_t = K_5 x \sigma_t^{1/m}$ is preferable (61-63) - Chapter 4.) Under multiaxial stress the equation is generalised as (64-66)

$$\bar{\sigma}_t = K \bar{\dot{\epsilon}}_t^{\,m}$$

where $\bar{\sigma}_t$ and $\bar{\dot{\epsilon}}_t$ are equivalent stress and strain rate such as are regularly encountered in continuum theories of plasticity (see, for instance, Hill (67)).

To satisfy the requirements of dimensional consistency equation 2.10 is occasionally written as (68,69)

$$\sigma_t = \sigma_o \left[\frac{\dot{\epsilon}_t}{\dot{\epsilon}_o}\right]^m \qquad .. \quad 2.11$$

This formulation, which reduces to equation 2.10 if the reference strain rate $\dot{\epsilon}_o$ is taken as unity, is not normally used.

Frequently terms are also included which account for the temperature and the grain size dependence. A commonly used general form (1) is

$$\dot{\epsilon}_t = \text{constant} \; \frac{\sigma_t^{1/m}}{L^a} \exp\,(-Q/kT) \qquad .. \quad 2.12$$

where L is the average grain size, a is a constant usually in the range 2-3 (see Chapter 3), Q is an activation energy, k is Boltzmann's constant and T is the absolute temperature of deformation. As with N and m, a and Q are also dependent on test and material parameters.

Another expression for isothermal deformation but incorporating grain size dependence is (70)

$$\dot{\epsilon}_t = \frac{A \; \sigma_t}{L^2} + B \sin h \; \sigma_t \qquad .. \quad 2.13$$

with A and B constants.

In addition two different rheological models have been presented (71,72) to describe superplastic flow. The first of these (71) yields equations similar to equation 2.6 but allows for the interaction of hardening and softening. The second (72) derives expressions dealing with variations in apparent viscosity and activation energy. More recently it has also been suggested (73,74) that under certain conditions the strain dependence of superplastic flow stress may not be negligible.

The empirical basis of the equations given in this Section will be described in Chapter 3.

Presently the stress-strain behaviour of superplastic materials (only structural superplasticity will be considered) will be examined and this is then followed by a rationalization of the relative stability of flow in terms of the above constitutive relationships.

2.3 The Deformation Behaviour of Superplastic Materials

2.3.1 The tensile stress-strain curve

The simplest and the most common method of tensile testing superplastic materials is to extend them in a machine whose cross-head moves at constant velocity. An initial load maximum and subsequent strain softening which often leads to a steady state are the common features of the load-extension curves obtained in these experiments (75-79). Although the load maximum has been attributed to the formation of the first neck (80,81) the very shallow nature of the latter precludes the possibility of the maximum from being solely due to necking (82). (Very rarely a wavy load-extension curve, with a relatively low amplitude of oscillation (83,84), has also been reported and the shape of the curve here has been traced to the shallow, multiple necks that form in the early stages of superplastic deformation.) A representative set of true stress-true strain curves obtained from constant crosshead speed tests is shown in Fig.2.1.

In a constant velocity (V) tensile test the true strain rate at time t is given by

$$\dot{\epsilon}_t = \frac{V}{l_o + Vt} = \frac{V}{l} \; . \qquad .. \; 2.14$$

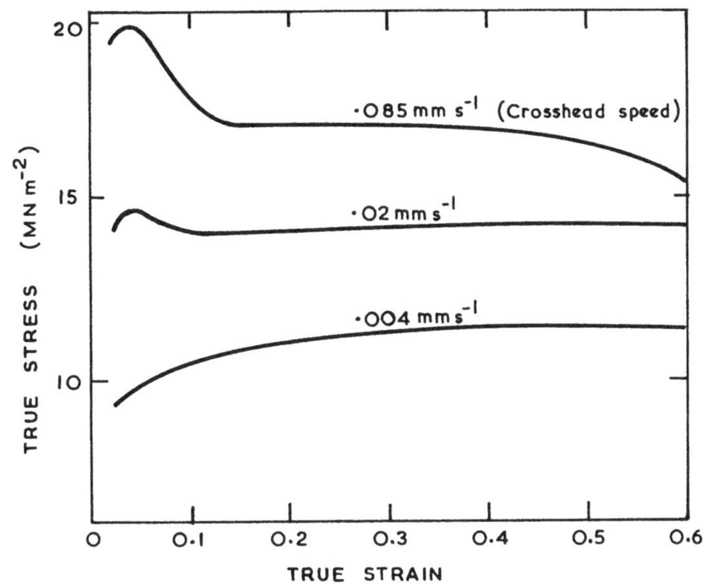

Figure 2.1 The effect of crosshead speed upon the true stress-true strain curve for superplastic commercial purity zinc. Specimen gauge length 63.5 mm (2.5 in.) (Naziri and Pearce (85))

Therefore, there is a gradual fall in strain rate with deformation. For instance, for the bottom curve of Fig.2.1 the true strain rate decreases from 1.2×10^{-3} to 8.1×10^{-4} s^{-1} as the true strain is increased from 0.1 to 0.5. For a significantly strain-rate sensitive material like the superplastic alloy this should lead to a measurable decrease in stress. Neglect of such variations has often led to the suggestion that the flow stress can be considered independent of strain only to a very limited degree. (Strain-rate compensated curves,however, are found to superimpose on results obtained from constant strain rate tests (81,86) - see below.) In tests with high initial strain rates ($\dot{\epsilon}_t \gtrsim 3 \times 10^{-2}$ s^{-1}) the fall off in stress is indeed very marked (87).

This inherent difficulty has led to the design of apparatus capable of carrying out constant true strain rate tests using either mechanical (88) or servocontrol systems (89,90). The use of this equipment requires

(i) that the actual length undergoing deformation is accurately known. This is normally taken as the gauge length. If deformation occurs within the grips this can lead to a systematic error which should be allowed for (90).

(ii) that the deformation is uniform till final necking. Actually, however, superplastic deformation proceeds with the formation of a series of diffuse necks (considered in detail in Section 2.5).

When necking is pronounced the constant strain rate control becomes invalid because of the localisation of deformation. Within these limitations constant true strain rate tests reveal little or no strain hardening (Fig.2.2). The apparent strain hardening seen in one

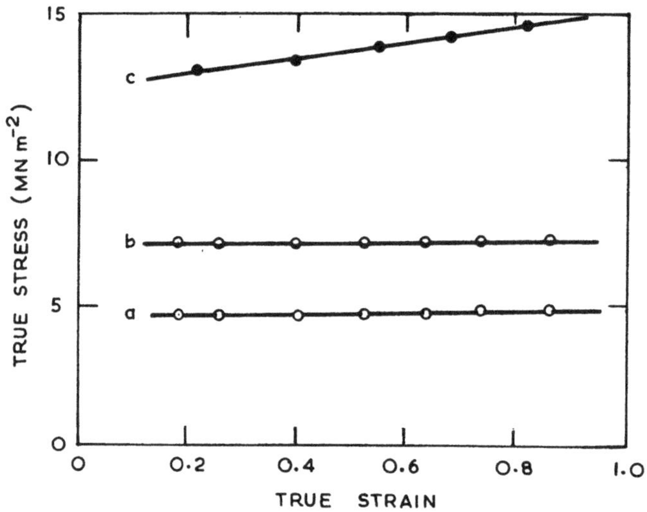

Figure 2.2 The true stress-true strain behaviour of superplastically deformed Sn-Pb eutectic tested in tension at constant true strain rate.

(a) $\dot{\epsilon}_t = 9.5x10^{-5}\ s^{-1}$; no elongation in the grips; test temperature 303K

(b) $\dot{\epsilon}_t = 1.9x10^{-4}\ s^{1}$; no elongation in grips; test temperature 303K

(c) $\dot{\epsilon}t = 2.0x10^{-4}\ s^{-1}$; 20% of specimen elongation in grips; test temperature 273.5K (Baudelet and Suery (90))

of the curves in Fig.2.2 has been attributed to deformation in the grips. In cases where substantial strain hardening is observed, it is the result of grain growth or phase coarsening (61,88,91-97).

Nevertheless, when the starting structure is not conducive to superplastic deformation, e.g. a cast or fibrous condition, grain refinement and/or morphological changes leading to strain softening take place (80,82,97). The flow stress, however, becomes strain independent once the initial unfavourable structure is fully destroyed (61). Thus subtle changes in microstructure can significantly alter the shape of the true stress-true strain curves (98).

The balance of evidence, therefore, suggests that strain hardening is observed when grain growth is present and strain softening when grain refinement and/or morphological changes take place. Strain independent behaviour is seen when the structure is fully stable. Very recently it has also been pointed out that as superplastic deformation is quasi-viscous, there is usually a delay time associated with the onset of steady-state and the stress-strain analysis is valid only beyond this point (99).

2.3.2 Tensile stress - strain rate variations

Steady state flow of structurally superplastic materials invariably shows an increase in flow stress with increasing strain rate. The variation of ln σ_t with ln $\dot{e}_t$ is sigmoidal and for convenience (80) the curve is divided into three regions (Fig.2.3 - schematic). The

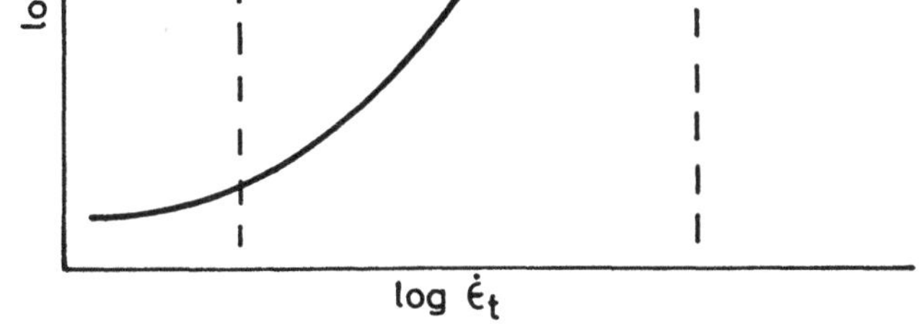

Figure 2.3 The relationship between stress and strain-rate for structurally superplastic materials (schematic).

region of maximum strain-rate sensitivity (region II) delineates the strain-rate range over which superplasticity occurs. At both low and high strain rates (regions I and III) the strain-rate sensitivity is small and these correspond to more conventional forms of plastic deformation. The strain-rate sensitivity of the flow stress in specific materials is considered in detail in the next chapter. For completeness at this stage, however, two examples of typical results are given in Fig. 2.4.

Assuming equation 2.10 to be valid it is possible to use the ln σ_t vs ln $\dot{\epsilon}_t$ curve to determine directly the strain-rate sensitivity index, although a range of other methods are also available (Section 2.7). There is some conflict over the strain dependence of such curves as shown in Fig. 2.4,e.g. there is evidence for strain independence (90) as well as for strain dependence (101).

More recent work (60,82) suggests that grain growth, whose importance changes with strain rate, can lead to a strain effect and a change in the shape of the stress-strain rate curves. Hysteresis present between stress-strain rate curves obtained with increasing and decreasing rates of deformation has also been primarily attributed to grain growth (102,103). Moreover, even when grain growth was present if the stress-strain rate curve was evaluated at (constant) strains greater than about 15-20%, the strain dependence was negligible.

2.3.3 Compressive behaviour

Compression testing of superplastic materials is far less widely reported than tensile testing although compressive behaviour is of considerable importance with regard to formability. Such results as are available (66,87,99,104,105) indicate that provided care is taken to eliminate or account for frictional effects then compression results can be superimposed on tensile data.

2.3.4 Torsion testing

As with compression testing it has been shown that the stress-strain rate curves derived from torsion tests (66,104) on the tin-lead eutectic and the zinc-aluminium eutectoid alloys are in agreement with those obtained using tensile tests.

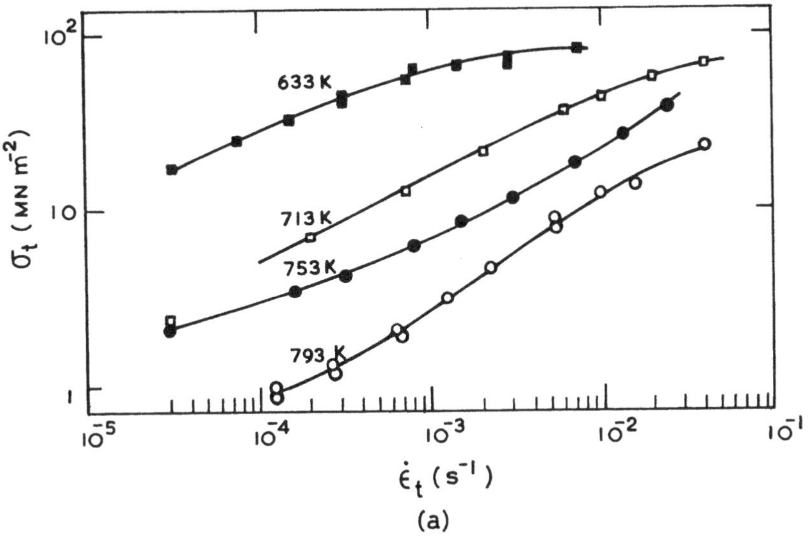

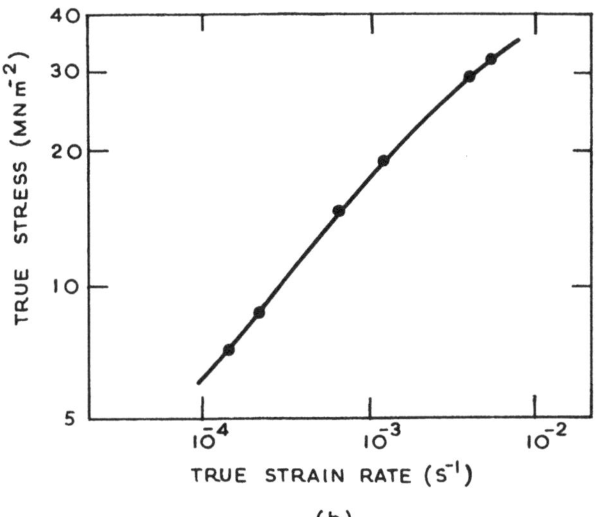

Figure 2.4 (a) True stress-true strain rate curves at different temperatures for the superplastic eutectic Al-$CuAl_2$, derived from constant crosshead speed tests (Holt and Backofen (100)).

(b) True stress-true strain rate curve at 298K for superplastic eutectic Sn-Pb determined using constant true strain rate apparatus (Baudelet and Suery (90)).

2.4 The Shape of the Deforming Specimen

Before studying necking in a tensile test it will be informative to examine the way in which a constitutive equation such as equation 2.5 influences the profile of a strained specimen. Assuming both strain-hardening and strain-rate sensitivity the specimen shape has been calculated by Stuewe (54) for increasing strain after employing for N and m (in equation 2.5) different values (Fig.2.5). In particular the

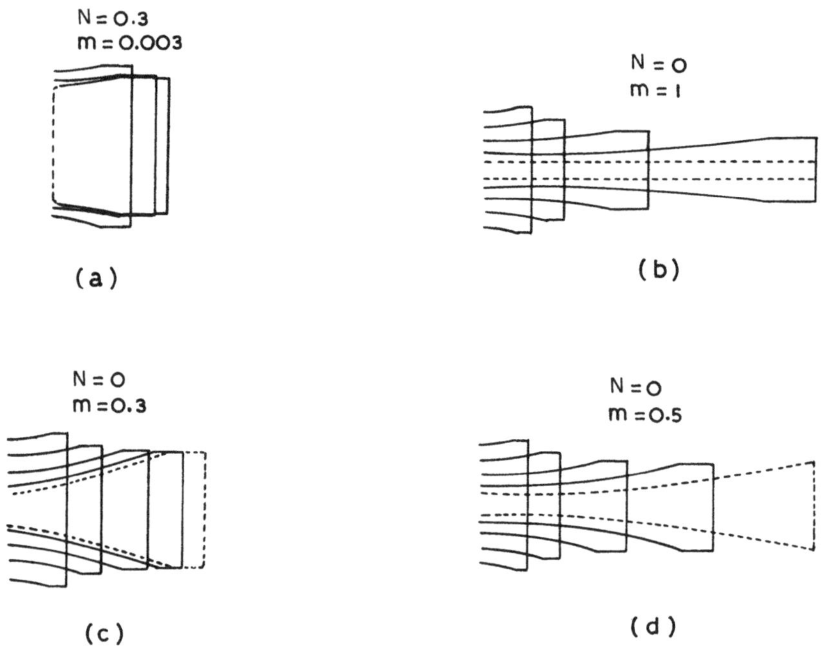

Figure 2.5 Calculated shape changes in tensile specimens deforming according to a flow law $\sigma_t = K\epsilon_t^N \dot{\epsilon}_t^m$

(a) strain-hardening dominant, N=0.3, m=0.003
(b) viscous flow, N=0, m=1
(c) weak rate sensitivity, N=0, m=0.3
(d) strong rate sensitivity (superplasticity), N=0, m=0.5 (Stuewe (54))

profiles of the rate-sensitive materials show the extreme elongations which can be achieved without necking.

The stable nature of superplastic deformation and the absence of other than diffuse necking are of considerable importance. Both the factors influencing the onset of necking and those determining the elongations need careful consideration.

2.5 Plastic Instability

2.5.1 The onset of necking

In a tensile test of a metal or alloy deforming under conditions where strain-hardening effects are dominant (m=0), e.g. deformation of copper at room temperature, the stress-strain behaviour is adequately described by equation 2.7. Here plastic deformation will be uniform and stable provided any localised decrease in cross-sectional area is offset by a corresponding increment in strain hardening (43). That is, for plastic stability

$$(\sigma_t + d\sigma_t)(A + dA) \geqslant \sigma_t A \qquad .. \quad 2.15$$

where $d\sigma_t$ and dA are the incremental changes in true stress and instantaneous cross-sectional area, respectively. When the second-order terms are neglected and constant volume deformation is assumed

$$\frac{d\sigma_t}{\sigma_t} \geqslant \frac{-dA}{A} = \frac{d\ell}{\ell} = d\epsilon_t \qquad .. \quad 2.16$$

where $d\ell$ is the incremental increase in length. Thus for stability

$$\frac{d\sigma_t}{d\epsilon_t} \geqslant \sigma_t \qquad .. \quad 2.17$$

Therefore from equation 2.7 the limit of uniform plastic flow is reached when $\epsilon_t = N$. Thereafter necking occurs and further deformation is concentrated in the necked region till failure takes place.

As $\dot{\epsilon}_t = \frac{d\epsilon_t}{dt} = \frac{-1}{A}\frac{dA}{dt} = \frac{1}{\ell}\frac{d\ell}{dt}$, for a linear (Newtonian) viscous, material obeying equation 2.8

$$\frac{-dA}{dt} = A\dot{\epsilon}_t = \frac{A\sigma_t}{K_3} = \frac{P}{K_3} \qquad .. \quad 2.18$$

where P is the load. Therefore for this type of material the rate of loss of area at any cross-section is dependent only on the load and is independent of the local cross-sectional area. Then a bar with a local shape discontinuity will reduce progressively with the shape being maintained and without significant local necking (Fig.2.6). This is, of course, the property that enables glass rods to be drawn in a flame, without breaking, to form fine filaments. It must be recognised, how-

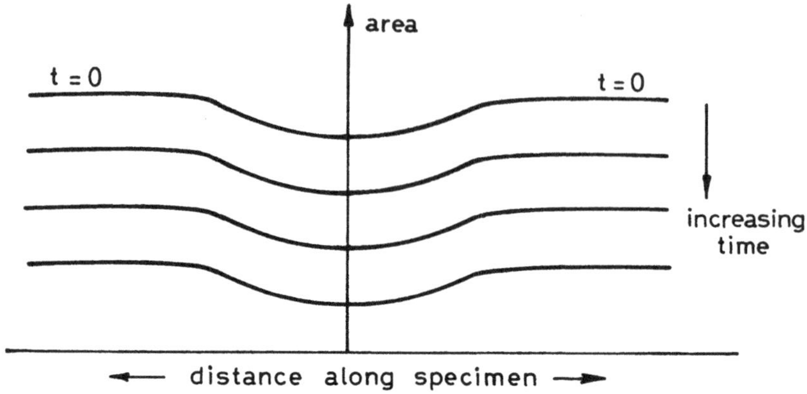

Figure 2.6 Distribution of areas in a tension bar of a viscous material

ever, that a loss of area at a constant rate means that smaller sections will reduce in area more rapidly than larger sections. Thus there will be a slow but steady enhancement of surface non-uniformities.

The more general case in which the materials exhibit both strain-hardening and rate-sensitivity has been considered by Rossard (51), Hart (52), Campbell (53), Klepaczko (106), Jonas et al. (107), Argon (108) and Hutchinson and Obrecht (109). For <u>isothermal</u> deformation, because of equation 2.4, this relation can be written in the differential form as

$$d\,\sigma_t = \left(\frac{\partial \sigma_t}{\partial \epsilon_t}\right) d\,\epsilon_t + \left(\frac{\partial \sigma_t}{\partial \dot{\epsilon}_t}\right) d\,\dot{\epsilon}_t \qquad .. \quad 2.19$$

For the most general case the partial derivatives should be replaced by coefficients which should be considered as material parameters that depend on specimen history (52). In principle, these parameters can also be deformation-dependent.

A stability criterion based on equation 2.19 would involve the study of two regions of a deforming specimen that differ in cross-sectional area by a small amount δA. The flow can be defined as stable if δA does not increase with deformation. Then, for stable deformation

$$\frac{\delta \dot{A}}{\delta A} \leq 0$$

where $\delta \dot{A}$ is the time derivative of δA. The stability criterion for this case becomes (52)

$$\left(\frac{\partial \sigma_t}{\partial \epsilon_t}\right) + \dot{\epsilon}_t \left(\frac{\partial \sigma_t}{\partial \dot{\epsilon}_t}\right) \geq \sigma_t \qquad .. \quad 2.20$$

On account of equation 2.5, equation 2.20 can be expressed in terms of the material parameters as,

$$\frac{N}{\epsilon_t} + m \geqslant 1. \qquad .. \quad 2.21$$

Evidently for the two special cases of zero rate dependence and strain-independent behaviour, in accordance with results presented earlier, inequality 2.21 reduces to $\epsilon_t \leqslant N$ (equation 2.17) and $m \geqslant 1$ (equation 2.18), respectively.

Rossard's (51) study concluded that for deformation at constant crosshead velocity there should be no necking when $m > 0.5$ and for deformation at constant true strain rate the requirement for the absence of necking was $m > 0.33$. These findings were confirmed by Hart (52) and it was also suggested that when the latter analysis is applied to superplastic flow the hardness parameter should be identified with the mean grain size (110). In contrast Campbell (53) established that there should be instability for $m \sim 0.5$. However, in this case the rate of increase of strain gradients would be very low.

As m for superplastic materials lies in the range 0.3 to 0.9 and $N \simeq o$, inequality 2.21 is not satisfied. Thus cross-sectional inhomogeneities are expected to grow slowly. From equation 2.10 and the relation $P = \sigma_t A$, however, it follows that

$$\dot{\epsilon}_t = -\frac{1}{A}\,(dA/dt) = (\sigma_t/K)^{1/m}$$

and $$(dA/dt) = -(P/K)^{1/m}\,(1/A^{(\overline{1-m}/m)}) \qquad .. \quad 2.22$$

As m increases from 0.3 to 0.9 the exponent $(1-m)/m$ decreases from 2.33 to 0.11. Clearly as m tends to unity the dependence of (dA/dt) on A is significantly diminished. Thus for superplastic alloys necks when present are diffuse and the extreme elongation is the result of the very high resistance to neck growth.

Necking strains predicted by the above linearized analyses do not agree well with experimental results. For instance, the strain at necking is equal to $N/(1-m)$ and N (same as predicted by Considere's construction), respectively, for the analyses of Hart (52) and Jonas et al. (107). However, the first measurable neck was encountered during superplastic flow at a strain more than an order of magnitude greater than that predicted by either criterion (81,82,97,111). It has, therefore, been concluded that Hart's analysis underestimates the effect of strain rate in postponing the onset of necking while Jonas et al. predict the necking strain to be strain rate independent (112). Both the analyses are also incapable of accounting for the postponement of necking with a decrease in the initial inhomogeneity of cross-section (112).

More recent stability criteria have been based on non-linear analyses. Troost and El-Magd (113) have considered only the rate-sensitivity of flow after assuming the existence of geometrical and structural inhomogeneities. With the aid of two adjustable parameters introduced in a non-linear fashion necking strains of the order of 100% have been predicted under certain conditions.

Non-linear analyses involving strain and strain-rate hardening as well as the existence of inhomogeneities in cross-section are also available (112,114-116). These calculations are essentially concerned with understanding the delayed appearance of <u>visible</u> necks during superplastic flow. Hutchinson and Neale (112) and Ghosh (116) assume the material response to be given by equation 2.5 while Ghosh and Ayres (114,115) employ an empirical constitutive equation

$$\sigma = K' \left[\epsilon^{N} + m \ln \left(\dot{\epsilon} / \dot{\epsilon}_{r} \right) \right] \qquad .. \quad 2.23$$

with K' a constant and $\dot{\epsilon}_r$ a reference strain rate at which no strain-rate hardening is present. The results of the different computations are somewhat similar and are therefore discussed together.

Hutchinson and Neale (112) define the 'critical state' for necking by the condition $d\epsilon_o/d\epsilon = 0$, where ϵ_o is the strain at any instant in the perfect region of the tensile specimen and ϵ is the strain at the same instant in the region of inhomogeniety. On the other hand, according to Ghosh (116) deformation is stable in the rising portion of the load-extension curve. Calculations reveal that neck growth is imperceptible at both the load maximum and a strain equal to $N/(1-m)$ (Hart's condition for onset of instability). This region of negligible neck growth has been described as 'quasi-stable'. The strain at which the first <u>measurable</u> neck forms, ϵ_o^c, is directly dependent on N and m but is inversely related to the size of the imperfection in cross-section, f. At low m, both m and N determine the magnitude of ϵ_o^c but when m is large the effect of N is rather small. When m $\gtrsim$ 0.6, ϵ_o^c of the order of 2.0 could be obtained (112,116).

When the constitutive equation is represented by equation 2.23, the predicted rates of neck growth are slightly faster and the instability strain decreases with an increase in strain rate. However, the latter has no effect on the growth behaviour of imperfections in cross-section (116).

It is relevant to note at this stage that the analysis of Hutchinson and Neale (112) is strictly valid only when m is small, $\epsilon_o^c < N$ and $m/N << 1$. Evidently these conditions are not satisfied for superplastic flow. Further, on account of multiple necking three dimensional effects could be important for superplastic flow while both the analyses of Hutchinson and Neale (112) and Ghosh (116) assume the existence of a uni-axial stress system. On the other hand, according to Langdon (117) a somewhat similar non-linear analysis (due to Nix and co-workers) is not very useful in predicting superplastic elongations.

Finally none of the analyses described in this section is capable of explaining the experimental observation that the onset of necking is postponed with a decrease in grain size (81).

2.5.2 The geometry of neck formation

The geometry of developing necks has been treated by a number of workers. The problem is one in which given an initial distribution of cross-sectional areas and a constitutive relation, it is necessary to predict the new distribution of areas at some arbitrary time during progressive elongation. The theoretical shape of the neck was calculated by Nadai and Manjoine (58), Hoff (118) and Campbell (53) for different conditions but the most significant step seems to be that of Wray (55) who made the distinction between failure by the development of a single neck and the formation of multiple necks. He recognized the existence of two types of plastic instability, which has since then been confirmed by other workers (81,97,119) for the case of superplastic flow. Type I instability involves the formation of several active necks in a specimen previously deforming uniformly (Fig.2.7). The second type of instability (Type II) involves the unstable situation that develops when a number of active necks are growing concurrently and one becomes dominant (Fig.2.8). This dominant neck then grows to produce failure.

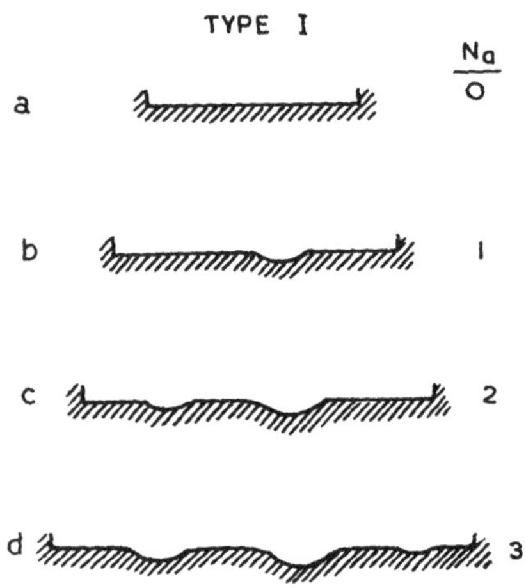

Figure 2.7 A schematic representation of Type I instability whereby active necks are formed (b-d) in an initially uniform specimen (a). N_a = no. of active necks (Wray (55))

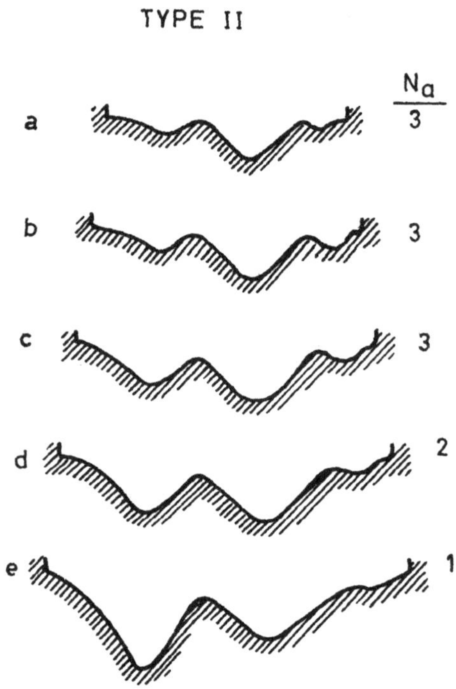

Figure 2.8 A schematic representation of Type II instability. In (a) the necks produced by earlier extension have approximately equal activity. Between (b) and (c) the neck on the left becomes dominant and eventually only this neck is active (d-e). N_a = no. of active necks (Wray (55))

During the period of domination the number of active necks decreases. It should be noted, however, that a decrease in the number of active necks does not unequivocally indicate the onset of Type II instability.

For some period of strain prior to the decrease being observed the activity in each neck may be different and the necks already unstable relative to each other. As yet there is no satisfactory theoretical treatment of Type II instability, although it has been speculated that dynamical effects will necessarily play a large part in the analysis (81, 97,119). Moreover, an experimental result which cannot easily be included in a theoretical frame-work is the decrease in the span between the two types of instability with increasing grain size or strain rate (81). It is likely that a successful treatment of Type II instability could explain the large elongations observed in superplastic alloys.

2.6 The Elongations of Rate-Sensitive Materials

The total elongation to failure will contain contributions from uniform deformation and from strain in any necked regions. For superplastic materials where multiple, diffuse necks form the latter contribution could be substantial.

If relatively homogeneous deformation is assumed, the elongation can be evaluated quantitatively (120) for a simple rate sensitive material obeying equation 2.10 by integrating equation 2.22. If a specimen has a uniform, initial area of cross-section A_o except for a locally reduced region whose area is αA_o and if due to deformation A_o reduces to A while αA_o reduces to βA, then from equation 2.22

$$\int_{t}^{t_o} (P/K)^{1/m}\, dt = \int_{A_o}^{A} A^{(\overline{1-m}/m)}\, dA = \int_{\alpha A_o}^{\beta A} A^{(\overline{1-m}/m)}\, dA \qquad .. \quad 2.24$$

or

$$A = \left[\frac{1-\alpha^{1/m}}{1-\beta^{1/m}}\right]^{m} A_o .$$

If the localisation of necking is ignored, then from definition

$$\%\ \text{elongation} = \left(\left[\frac{1-\beta^{1/m}}{1-\alpha^{1/m}}\right]^{m} - 1\right) 100 \qquad .. \quad 2.25$$

If it is assumed that failure occurs by rupture of the cross-section of the smallest area, i.e. $\beta = o$, then equation 2.25 becomes (121)

$$\%\ \text{elongation} = e_o + 100\left[(1-\alpha^{1/m})^{-m} - 1\right] \qquad .. \quad 2.26$$

where e_o is the elongation for m=0. Taking α as 0.99 and e_o to be 40%, a typical tensile elongation for a rate insensitive material, the curve of Fig.2.9 is derived for different m values. The strong depen-

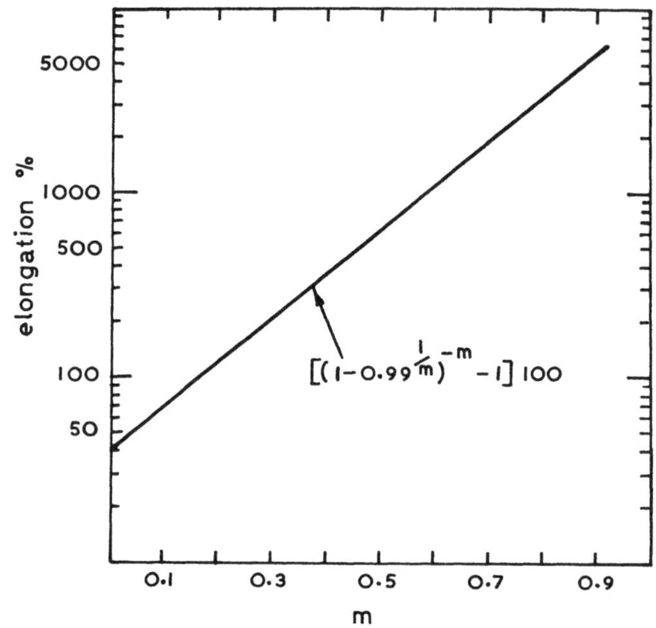

Figure 2.9 The relation between elongation and strain-rate sensitivity index calculated using equation 2.26

dence of elongation on m is clearly apparent. However, it is relevant to note at this stage that this analysis which ignores local deformation and assumes m to be a constant is idealised so far as superplastic flow is concerned because (a) multiple necks of varying activity are present right from the early stages of deformation (Section 2.5.2) and (b) m is dependent on the experimental variables like temperature, grain size, strain rate etc. (Chapter 3). Moreover, although equation 2.26 predicted the correct trend, numerical fit was not satisfactory.

In general a direct relation between the value of m and the elongation exists (for recent experimental verification see, for example, refs. 111,122-124). The experimental data verifying the relationship was collated at an early stage by Woodford (125) but as can be seen from Fig. 2.10 there is considerable scatter. For instance,

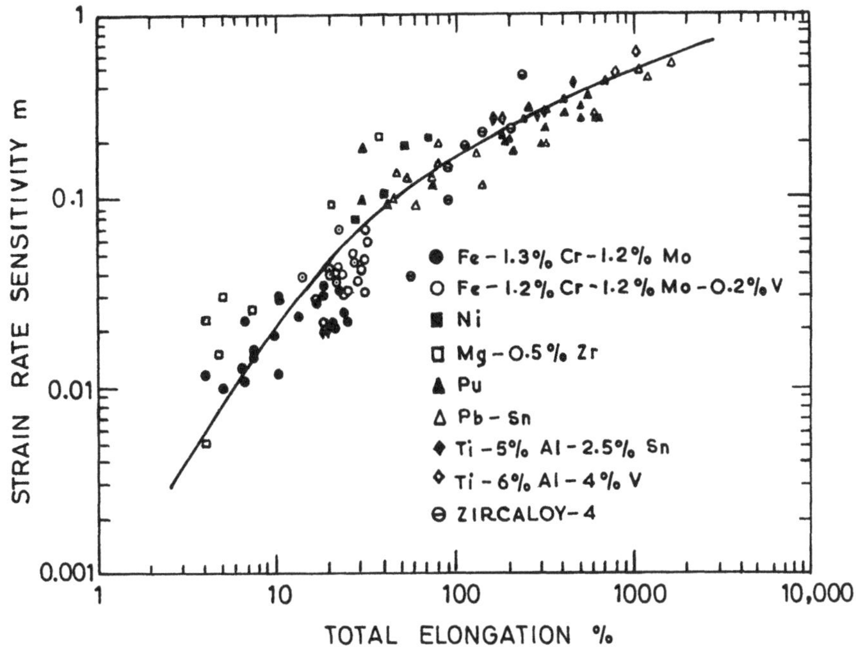

Figure 2.10 Correlation between strain-rate sensitivity and total elongation for a variety of materials (Woodford (125))

maximum m recorded in case of Al-Si eutectic, C-Mn low alloy steel and Al-Cu eutectic is about 0.7. The maximum elongation obtained in each case, however, is vastly different. On the other hand, the maximum m for the Zn-Al eutectoid is typically of the order of 0.5. But the maximum elongation for this alloy is much greater than that reported for the other three cases (80,126,127). Moreover, in some cases the maximum elongation is encountered at a strain rate an order of magnitude greater than that for which m is a maximum (111,128). These anomalies could be attributed at least partly to (a) the change in m with deformation in a constant velocity test through a change in strain rate, (b) deformation-induced grain growth which makes m strain dependent, (c) the differences in specimen size and geometry, (d) the variations in the accuracy of machining of specimens and (e) the differences in the number and distribution of cavity nucleating sites between specimens which, in alloys that are prone to cavitation, significantly alter the cavitation rate and hence the total elongation (80,103,119,122,125,129).

It should further be observed that in non-superplastic materials also the elongation to failure cannot be considered as the most reliable measure of ductility and that in case of superplastic flow as yet there are no data relating elongation and m for controlled tests at constant true strain rate.

Differences in specimen geometry can be responsible for differences in the relative contribution of necked regions. The extent to which specimen geometry affects superplastic elongation has been studied by Morrison (130). Firstly, it was shown that the elongations of different regions of a superplastically deforming alloy (tin-lead) specimen were far from uniform (Fig.2.11). Secondly, an empirical relationship

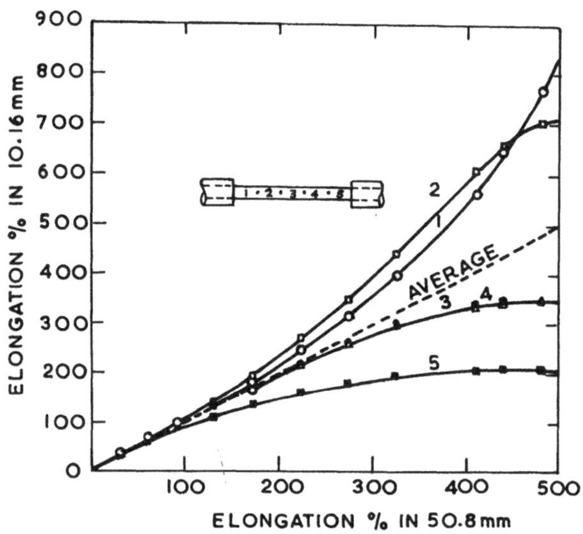

Figure 2.11 Elongations measured in different zones of a gauge length subdivided into five parts. Lead-tin alloy, total elongation 500 percent (Morrison (130))

$$\% \text{ elongation} = B\, m^2\, (d_o / \ell_o)\, 100 \qquad .. \quad 2.27$$

where B is a constant, d_o is the initial specimen diameter and ℓ_o is the original gauge length, was derived. One of the main difficulties associated with the use of equation 2.27 is the choice of value for m which, as discussed earlier, could vary in a tensile test. There is some support for the view that the minimum value of m is most appropriate.

Two other empirical relationships for the total elongation are also occasionally found in the superplasticity literature. According to Al-Naib and Duncan (65)

$$\% \text{ elongation} = (100^{m} - 1) \times 100 \qquad .. \ 2.28$$

while Ghosh and Ayres (114,115) have employed the equation

$$\text{True strain at fracture} = - m \ln (1 - \alpha^{1/m}) \qquad .. \ 2.29$$

with m equal to the value of the strain-rate sensitivity index just before fracture ('terminal' value).

At the moment the expressions available for predicting elongation to fracture do not take into account the dependence of m on temperature, strain rate and structure. Nor do they account for the fact that even when m is approximately constant, elongation is affected by changes in composition, test temperature, strain rate and structure (both initial and final grain sizes) (60,81,111,117,124,127,131).

2.7 The Strain-rate Sensitivity Index, m

Throughout this Chapter it has been clear that the strain-rate sensitivity index is a very important parameter in characterising structural superplastic deformation. If equation 2.10 is obeyed,

$$m = \frac{d\ (\ln \sigma_t)}{d\ (\ln \dot{\epsilon}_t)} \qquad .. \ 2.30$$

This is the slope of the $\ln \sigma_t - \ln \dot{\epsilon}_t$ curve (Fig.2.3). There are various methods of measuring m but broadly all lead to a variation of m with $\dot{\epsilon}_t$ of the form shown in Fig.2.12. m also varies with temperature and structure (grain size).

2.7.1 Determination of m from the $\sigma_t - \dot{\epsilon}_t$ curve

In this method m is determined as the slope of the experimental $\ln \sigma_t - \ln \dot{\epsilon}_t$ plots. The latter are obtained using instantaneous values measured in the steady state region (the onset of steady state is often defined at the load maximum in the load-extension plots (Section 2.3). However, when strain softening, which is a function of strain rate, is significant the steady state is reached at a strain beyond which there is no sharp decrease in stress with a small increase

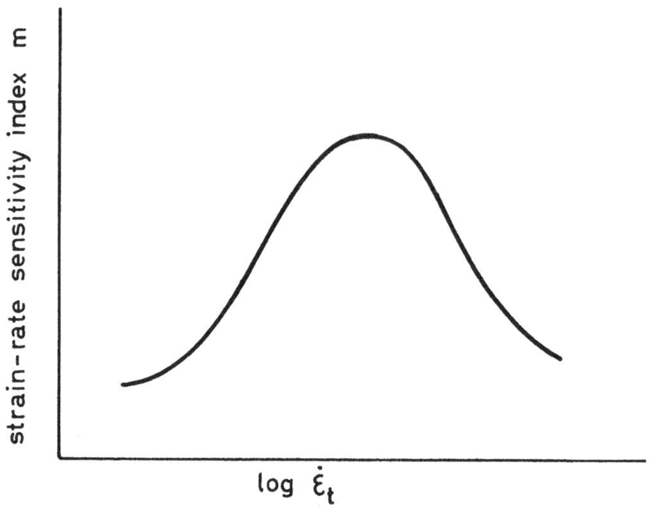

Figure 2.12 The variation of the strain-rate sensitivity index with $\dot{\epsilon}_t$ (schematic)

in strain and equation 2.10 is valid) of the load-elongation curves or by employing constant true strain-rate deformation. The rate at which data are collected can be increased if incremental increases in the strain rate are carried out in a single test. The slope of the ln σ_t - ln $\dot{\epsilon}_t$ plot (=m) is measured either graphically or better by using a curve-fitting procedure and carrying out differentiation (132). This method of m evaluation has been used successfully for a variety of materials (56,75,76,90,95,133,134). With constant true strain-rate deformation the measured value of m was found to be independent of strain or time (90).

2.7.2 Determination of m using change in strain-rate method

This is the most widely used procedure and is normally based on the description of Backofen et al. (42). (The actual technique, however, can be traced to Haigh (16,135).) As shown in Fig.2.13 if the crosshead velocity is suddenly increased from V_1 to V_2 there is a corresponding increase in load. If straining is continued for a few percent to eliminate transient effects a load comparison can be made. The lower velocity curve is extrapolated to establish a common strain for measurement (points A and B in Fig.2.13 are of equal strains). If m is assumed to be nearly independent of strain rate in the range covered by the velo-

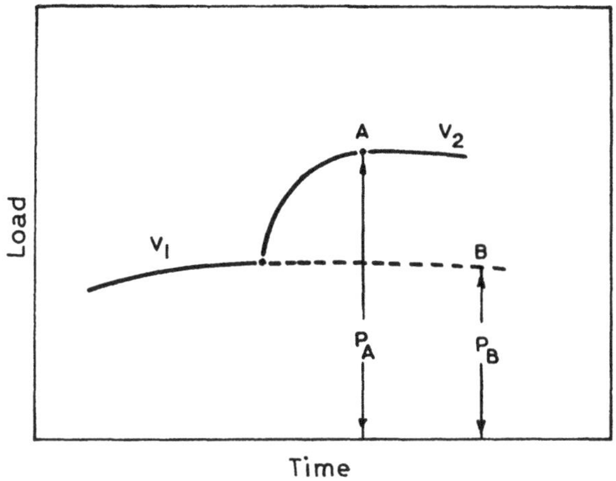

Figure 2.13 A schematic load-time diagram representing a velocity change from V_1 to V_2. Times A and B represent the same strain at the different pulling speeds (Backofen, Turner and Avery (42))

city increase, then

$$m = \frac{\ln\ (P_A\ /\ P_B)}{\ln\ (V_2\ /\ V_1)} \qquad .. \quad 2.31$$

Typical results using this procedure can be found in the papers of Backofen et al. (42), Holt and Backofen (100), Lee (136) (constant crosshead velocity tests); Herriot et al. (93) and Suery and Baudelet (82) (constant true strain-rate tests).

Variations to this method can be introduced by identifying the critical points on the load-time curve slightly differently (92,137). Considering Fig.2.14, a more detailed version of Fig.2.13, the alternatives can be examined. As before, equation 2.31 gives the m values for the method of Backofen et al. (42). Another approach (85,101,130) is to calculate the true stresses and the true strain rate at the points of maximum load (A and C in Fig.2.14). Then,

$$m = \frac{\ln\ (\sigma_t^A\ /\ \sigma_t^C)}{\ln\ (\dot{\epsilon}_t^A\ /\ \dot{\epsilon}_t^C)} \qquad .. \quad 2.32$$

(This is a variation on the method referred to in the previous section. In this method the stresses are evaluated at different strains while equation 2.10 is strictly valid at constant strain.) According to

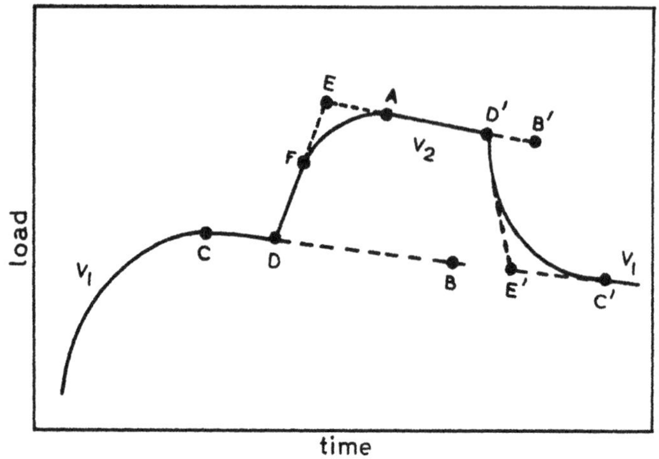

Figure 2.14 A detailed load-time curve for velocity cycling from V_1 to V_2 and vice-versa

Hedworth and Stowell (92) in case of the Al-Cu eutectic corresponding to two velocity changes equation 2.31 yielded values of $m = 0.67 \pm 0.01$ and $m = 0.72 \pm 0.01$ while equation 2.32 gave $m = 0.96 \pm 0.02$ and $m = 1.25 \pm 0.02$. The discrepancy was explained by recognising that the strain rate was changing throughout the test. In addition as the maximum in the load-extension curve can precede the onset of steady state (see previous section) work-hardening transient effects can also influence the value of m when equation 2.32 is used (137).

A velocity-change test can also involve a velocity decrease (92) for which equivalent forms of equations 2.31 and 2.32 can be defined. With reference to Fig. 2.14 these are,

$$m = \frac{\ln (P_{C'}/P_{B'})}{\ln (V_1/V_2)} \quad .. \quad 2.33$$

and

$$m = \frac{\ln (\sigma_t^{C'}/\sigma_t^{A})}{\ln (\dot{\epsilon}_t^{C'}/\dot{\epsilon}_t^{A})} \quad .. \quad 2.34$$

respectively.

Tests were carried out in which six strain-rate increments and five strain-rate decrements were applied successively. Equations 2.31

and 2.33 gave m = 0.69 ± 0.02 and m = 0.64 ± 0.02, respectively. The discrepancies observed between equations 2.32 and 2.34 were reported to be large.

Previous to this detailed study it had been reported (87) that different m-values were obtained for velocity increases as compared with velocity decreases. On the other hand, a subsequent study (138) showed no dependence of m-value on the direction of velocity change.

As a result of the discrepancies observed, another alternative method of measuring m was proposed (92). This method involved the extrapolation of the slowly changing (or approximately constant) part of the load-time curve back to the instant the velocity change was made. m was then defined as (see Fig. 2.14)

$$m = \frac{\ln\left(\sigma_t^E / \sigma_t^D\right)}{\ln\left(\dot{\epsilon}_t^E / \dot{\epsilon}_t^D\right)} \quad \text{or} \quad = \frac{\ln\left(\sigma_t^{D'} / \sigma_t^{E'}\right)}{\ln\left(\dot{\epsilon}_t^{D'} / \dot{\epsilon}_t^{E'}\right)} \qquad \text{.. 2.35}$$

Results obtained with the aluminium-copper eutectic using equation 2.35 yielded m = 0.71 ± 0.02 for velocity increments and m = 0.69 ± 0.01 for velocity decrements. Consequently this latter method was recommended, particularly because of its simplicity. The above consistency of results is largely because

$$\ln\left(\dot{\epsilon}_t^E / \dot{\epsilon}_t^D\right) = \ln\left(\dot{\epsilon}_t^{D'} / \dot{\epsilon}_t^{E'}\right) = \ln\left(V_2/V_1\right).$$

Two other variations have also been recently discussed (129, 133) (Fig. 2.14):

$$m = \frac{\ln\left(\sigma_t^A / \sigma_t^D\right)}{\ln\left(\dot{\epsilon}_t^A / \dot{\epsilon}_t^D\right)} \qquad \text{.. 2.36}$$

and

$$m = \frac{\ln\left(P_F/P_D\right)}{\ln\left(V_2/V_1\right)} \qquad \text{.. 2.37}$$

In the first method the stresses are determined at different strains and the stress at the higher velocity corresponds to the load maximum while that at the lower speed pertains to the instant of velocity change, i.e. beyond the load maximum. The second method, on the other hand, is of dubious significance because (a) the absence of machine effects and rapid recorder response are essential for locating F accurately and (b) in the range F to A stress increases with a decrease in strain rate and therefore equation 2.10, the basis of m determination, is not obeyed.

2.7.3 Determination of m from stress-relaxation tests

The stress-relaxation technique has been widely used to study the time-dependent plastic flow of crystalline solids (see, for instance, Feltham (139) and Guiu and Pratt (140)). The procedure involves the plastic deformation of the solid to some chosen stress level at which stage the crosshead movement is stopped and a continuous decrease in stress is observed as a function of time, t (Fig. 2.15). Strain occurs at

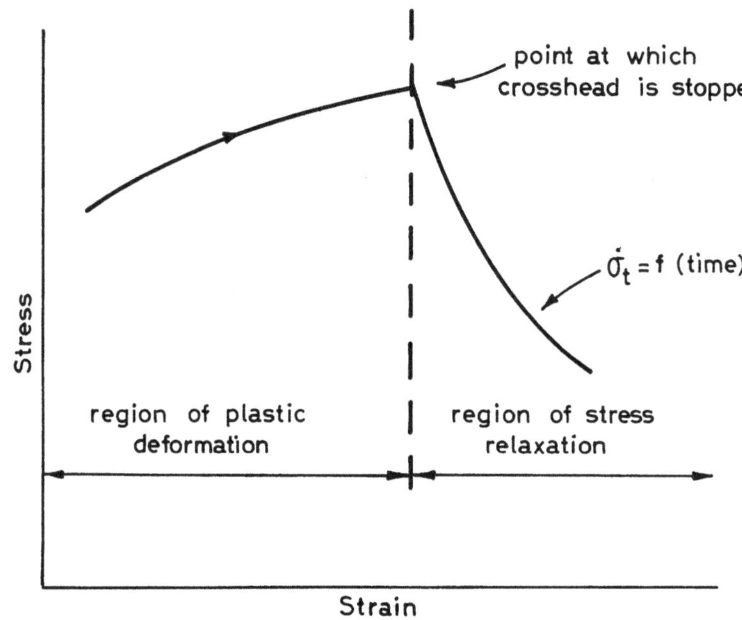

Figure 2.15 A typical stress-relaxation curve obtained by arresting the testing machine crosshead (schematic)

this stage as the elastic strain in the machine and specimen is relieved.

In analysing the results of stress-relaxation tests there are three approaches.

(i) Feltham (139) assumed a relation similar to equation 2.10 with m independent of strain rate over the range of interest. For this case it was shown that

$$\ln \sigma_t = C + \frac{m}{(m-1)} \ln (t + D) \qquad .. \; 2.38$$

where C and D are constants. Thus a plot of $\ln \sigma_t$ against $\ln t$ should yield a straight line of slope $m/(m-1)$ for $t \gg D$.

(ii) An alternative approach (140) starts with the same assumptions as above and leads to the relation

$$\frac{1}{m} = \frac{d(\ln - \dot{\sigma}_t)}{d(\ln \sigma_t)} \qquad .. \; 2.39$$

where $\dot{\sigma}_t$ is the rate of relaxation of stress. Thus m can be found from the slope of a plot of $\ln (-\dot{\sigma}_t)$ against $\ln \sigma_t$. For this procedure it is not a necessary requirement that m is constant.

(iii) A third method is based on a constitutive equation which assumes the existence of an internal stress. That is,

$$\sigma^* = \sigma_t - \sigma_i = K^* \dot{\epsilon}^{m^*} \qquad .. \; 2.40$$

Then m* can be obtained as the slope of a $((d \ln(-\dot{\sigma}_t))/d\sigma_t)^{-1}$ vs σ^* plot (92,129). The relation between m and m* is given by (129)

$$m = m^* (1 - (\sigma_i / \sigma_t)) \qquad .. \; 2.41$$

Stress-relaxation tests, mostly by using equation 2.39, have been carried out on a wide range of structurally superplastic materials (91,92,141-143). In most cases it has been reported (141-143) that the value of m measured from stress-relaxation tests agrees within experimental error with that determined using a method such as that of Backofen et al. (equation 2.31). Hedworth and Stowell (92), however, observed that for the Al-$CuAl_2$ eutectic stress-relaxation testing gave $m = 0.44 \pm 0.01$ under conditions where change in strain-rate testing yielded $m = 0.70 \pm 0.02$.

2.7.4 Comparison of m values determined using different techniques

Apart from Hedworth and Stowell (92) (see above) a number of investigators have used more than one procedure to determine m and have compared the results. According to Hildebrand, Michalzik and Simmon (134) the method given in Section 2.7.1 (from the slope of a $\ln \sigma_t - \ln \dot{\epsilon}_t$ plot) and that of Backofen et al. (equation 2.31) resulted in similar values of m. Elaborating on this point Nuttall (137) has remarked that the values of m for the two methods were equal only when the flow stress was independent of strain. However, when linear work hardening occurred (often due to grain growth) the method of Backofen et al. gave a lower value in an increasing strain rate test but a higher figure for a decrease in strain rate. The more recent observations (56,90,95) that (a) the peak m values obtained by the method of Section 2.7.1 were greater than those calculated from a change in strain rate and (b) the peak occurred at a strain rate approximately an order of magnitude greater (Fig.2.16) in case of the former confirm the above suggestion

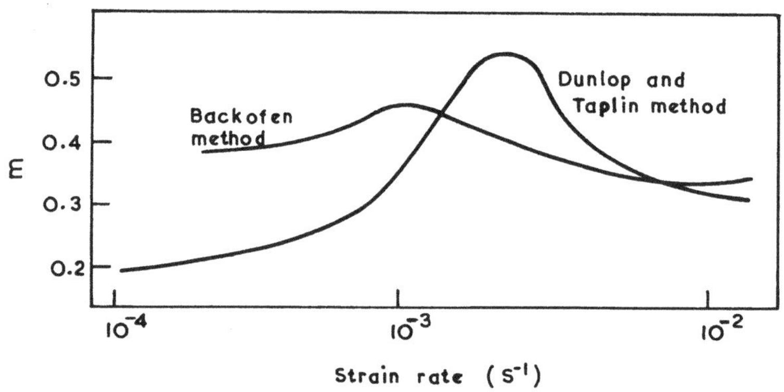

Figure 2.16 m-strain rate curve for superplastic aluminium-bronze obtained using
(i) change in strain-rate (Backofen) method,
(ii) slope of $\sigma_t - \dot{\epsilon}_t$ curve (Dunlop and Taplin) method
(Ducheyne and De Meester (95))

(of Nuttall).

The change in strain rate test and the stress relaxation method, on the other hand, often give similar values of m (123) - see also Fig.2.17.

After a detailed comparison, Edington et al. (129) have concluded that all methods except the one using equation 2.37 resulted in similar values of m. Greatest consistency, however, was obtained with equation 2.35 (for reason see previous section) although almost as good consistency was present in case of the more popular procedure based on equation 2.31.

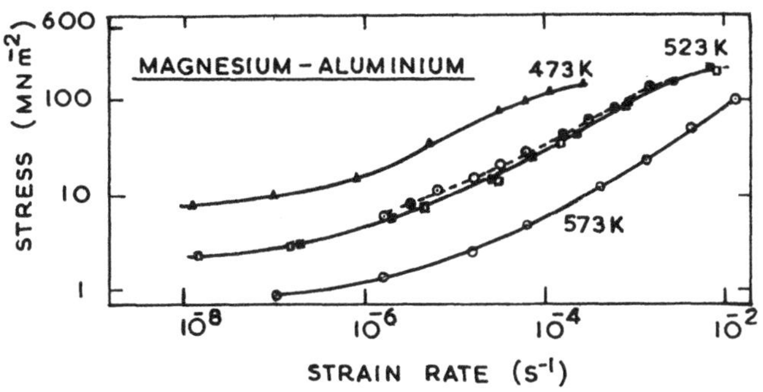

Figure 2.17 Stress-strain rate relationships for the superplastic magnesium-aluminium eutectic determined using stress-relaxation (full curves) and change in strainrate (dashed curve) procedures (Lee and Hart (141))

2.7.5 The physical significance of m value measurements

The dissimilarities among the various methods as well as their relevance to fundamental studies have been commented upon by a number of authors (60,92,95,129,133,144). The disagreement in the m values reported by the various techniques could largely be attributed to (a) necking, which if significant can cause a strain dependence of m, (b) the differences in primitive defect structure arising from the different magnitudes of strain employed in each procedure, (c) a wide variation in the value of m over the strain rate range covered by a given method, (d) grain growth which has a direct relation to strain and an inverse dependence on strain rate and (e) the sign and the magnitude of the strain rate change involved.

Ducheyne and De Meester (95) and Arieli and Rosen (133) have recommended the adoption of the method of Section 2.7.1 for m value measurements when there is negligible grain growth during superplastic deformation. This is because the procedure involves no assumptions or extrapolations and compares similar initial structures at all strain rates. For best results, however, a new specimen should be used at each strain rate and the method is time-consuming. Moreover, in this method m is derived after lengthy calculations and it cannot be directly obtained by using a simple equation.

Hedworth and Stowell (92), on the other hand, have concluded that the stress relaxation tests yielded the most satisfactory fundamental data because of the very small strains involved and the consequent constant structure conditions. It was felt that in this test both grain coarsening and changes in the primitive defect structure would be avoided.

More recently, however, this method has come in for criticism (129,145-147). At elevated temperatures stress-strain rate evaluation from stress relaxation tests is not accurate as the machine stiffness is not constant and numerous machine and instrumentation problems exist. The initial portion of the load-extension curve, on the other hand, could contain a plastic flow component while for the interpretation of data this portion is assumed to be elastic. Moreover, the apparent viscosity' of the grain boundaries leads to a time-dependent release of elastic energy which may appear as an internal stress. Finally, as much of the deformation is concentrated at grain or interphase boundaries (see Chapters 3 and 4) where diffusivities are high, rapid structural modifications during relaxation are inevitable.

In view of the above, the method of Backofen et al. (equation 2.31) has consistently found favour (see, for example, refs. 92,129 and 137) although it employs a greater strain before m evaluation. The consistent values obtained for m and the simplicity of the test procedure are the main attractions. The two sources of error, viz. (i) the extrapolation of the load-time curve at the lower crosshead speed and (ii) the change in the crosshead speed to chart speed ratio arising from the velocity change are, on the other hand, considered to be of a minor nature (92,137). Thus this method could profitably be used to compare and characterise different structurally superplastic materials, particularly after ensuring that (a) the strain applied prior to the velocity change has as low a value as is consistent with the attainment of steady state (to eliminate necking and grain growth) and (b) the (V_2/V_1) ratio is only slightly greater than unity (usually about two) so that m is relatively constant in this range.

2.8 Strain-rate Sensitivity in Environmentally Superplastic Materials

The evaluation of the strain-rate sensitivity during superplasticity produced by cycling through a phase transformation has been described by Kot and Weiss (148) and is based on the earlier work of Oelschlaegel and Weiss (29). The method is very similar to that described in Section 2.7.1 for structurally superplastic materials. Specimens under constant load were cycled through a phase transformation (Fig.2.18) and temperature-time and extension-time curves obtained.

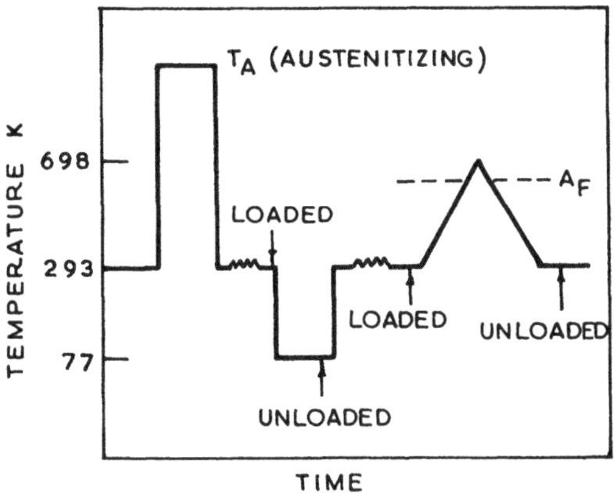

Figure 2.18 Schematic temperature-time diagram showing the experimental procedure for cycling an Fe-32.9% Ni alloy under constant load (Kot and Weiss (148))

The plastic flow per cycle was determined for both heating and cooling. A constitutive relation as given in equation 2.10 was assumed. The strain-rate was evaluated by dividing the plastic elongation per cycle by the cycle time. Since the stress was known, a ln σ_t-ln $\dot{\epsilon}_t$ curve could then be plotted and the strain-rate sensitivity index was found by determining the instantaneous slope (that is, using equation 2.30). In both studies (29,148) the m-value was reported to be $\sim$1.0.

This approach, in contrast to isothermal studies of structural superplasticity, requires that K in equation 2.10 is independent of temperature. Other errors can result from the neglect of the high internal stress field caused by the proceeding transformation (149,150) and from the assumption that the externally recorded strain-rate is representative of events in the vicinity of the transformation interface. As a result it is probable that both the stress and the strain-rate are significantly underestimated. The physical meaning of m as determined by the above procedures is therefore not clear.

This difficulty has been further examined by Yoder and Weiss (151). Using a variant of the strain-rate change procedure (cf. Section 2.7.2) during the period of transformation they recorded m-values of about 0.35. This was of the same magnitude as the m-value determined in the untransformed α-Fe_3C aggregate undergoing structurally superplastic deformation. This result led to the conclusion that the strain-rate sensitivity was not of importance in stabilising flow during transformation superplasticity but that stability resulted from normal strain hardening effects. This is considered further in Chapter 5.

2.9 Other Test Procedures for Studying Structural Superplasticity

2.9.1 Indentation tests

Indentation hardness testing is widely used as a means of the non-destructive assessment of plastic properties. Assuming that creep during indentation hardness testing is analogous to steady state creep, Murty et al. (152) have described a simple way of determining the stress exponent (1/m in equation 2.12) and the activation energy for deformation. In case of the tin-lead eutectic, values of these quantities obtained by using standard Brinell hardness tests were similar to those obtained from tensile tests.

The time dependence of indentation size is a well-known property of hardness tests on superplastic materials. By defining the 'superplasticity index' as the numerical difference in dial readings noted at two fixed times following the application of the major load in case of the Rockwell hardness test, Marcantonio et al. (153) have described a rapid and simple empirical procedure which is suitable for adoption as a control test for production or acceptance of structurally superplastic materials. Belk (154), on the other hand, has pointed out that the method has to be standardized for each alloy and that there is no obvious way of correlating the superplasticity index with the more common strain-rate sensitivity index. As the diameter of indentation is affected much less by elastic recovery compared with the depth of penetration, he advocates the measurement of the former when using this technique as a control test. Modification of other machines, e.g. the Vickers hardness machine, extension of the method to characterise superplastic behaviour at elevated temperatures and evaluation of m by following a procedure similar to the one outlined by Murty et al. (see above) are the advantages claimed in favour of this modification (which, however, are yet to be demonstrated).

2.9.2 Measurement of internal stresses

Methods for the measurement of internal stresses in superplastic materials are based on those developed in studies of high temperature creep (155-157). Measurements of the internal stress assume the existence of a constitutive relation such as that given in equation 2.6. Two common methods are used, (i) the stress reduction technique, in which the applied stress is suddenly reduced during tensile testing and a stress relaxation test is carried out from the reduced stress level. If this reduced stress level is below the internal stress reloading occurs while if it is above unloading is seen. The internal stress is the reduced stress at which the unloading rate immediately following stress reduction is zero. (ii) the stress relaxation technique, in which a normal stress-relaxation test (see Fig. 2.15) is carried out without stress reduction. Here the internal stress is taken as the stress at which the unloading rate tends to zero.

The existence of an internal stress has been claimed in case of some superplastic alloys (56,145,158) and this point is considered further in Chapters 3 and 4.

2.9.3 Measurement of grain boundary shear

Grain boundary sliding is considered to be an important mechanism of deformation in case of structural superplasticity (see Chapters 3 and 4). Measurements of the contribution of grain boundary shear to the total strain have been frequently carried out in high temperature creep studies and the different methods available are discussed by Stevens (159). Hazzledine and Newbury (61) and Geckinli and Barrett (146) have, on the other hand, questioned the significance of the numerical values obtained for the contribution from grain boundary sliding(when these techniques are applied to superplastic deformation) because (a) surface measurements may not be representative of bulk behaviour and (b) as (grain) neighbours change during superplastic flow (see Chapter 3) measured boundary offsets may not be very meaningful. Nevertheless, the methods have been adopted for use with superplastic materials, see for example, Holt (160) and Vatsava and Langdon (161).

2.9.4 Damping characteristics

The damping capacity of superplastic zinc-aluminium eutectoid alloy has been examined by Nuttall (162) using torsional free decay. Both the logarithmic decrement and the specific damping capacity were determined. However, the high damping capacity observed in the superplastic condition made the interpretation of results difficult and this limits the fundamental usefulness of this test procedure.

2.9.5 Miscellaneous tests

As the behaviour of superplastic materials during sheet metal forming is likely to be different from their tensile response, Belk (163) has suggested a method in which an equivalent of K (denoted as $\hat{K}$) in equation 2.10 is calculated from the vacuum forming time, which is the time taken at a given temperature by a disc of the material of a particular thickness to deform in a circular jig under one atmosphere pressure to a pre-determined geometric shape. Superplastic characteristics are inversely related to the value of $\hat{K}$.

As friction plays an important part in superplastic forming with the aid of compressive forces, there is also a need to understand work piece-tooling interactions in case of strain-rate sensitive materials (164). The ring test in which a thin ring of the superplastic material is compressed between the tools, with appropriate lubrication, to a pre-determined reduction in thickness has been employed for this purpose. The inner diameter of the ring for a series of thickness reductions is then plotted against the reduction. With the help of a calibration chart specifically developed for strain-rate sensitive materials good estimates of the friction factor are claimed to have been obtained. However, it must be noted that the constitutive equation used in the analysis, viz. $\dot{\bar{\epsilon}}_t = K_1(\bar{\sigma}_t - \sigma_o) + K_2\bar{\sigma}_t^N$, is somewhat uncommon.

3. Structural Superplasticity – Experimental

In Chapter 1 the historical aspects of studies in superplasticity were discussed and the Chapter concluded with a description of the revival of interest in structural superplasticity following developments in 1962 and 1964. In this chapter the contemporary experimental evidence for the many characteristics of structural superplasticity will be examined. Attention will be concentrated on work carried out subsequent to the dates given above. In some special cases it may be necessary to make reference to early work but this should only be so when unique alloys have been used or when observations have been made which have not been since repeated. In addition, it was not considered appropriate to include in this chapter much of the Russian work not available in translation. The theoretical explanations which have been put forward to account for structural superplasticity will be considered in Chapter 4.

An extremely wide range of metals and alloys have been reported as exhibiting anomalous ductility or extreme elongations which qualify them to be considered as structurally superplastic. These are materials in which a stable ultrafine grain size exists. Table 3.1 presents a comprehensive list of these materials giving values for the strain-rate sensitivity index and the elongation. This Table is arranged broadly in a sequence of increasing temperature range of superplastic behaviour subject to two constraints, viz. (a) alloys based on a particular metal or commercial classification are grouped together and (b) for a given base element the alloys are arranged in increasing complexity of composition. Examination of this Table shows that it contains 126 metals and alloys. The lowest temperature of superplasticity is near room temperature (293 K) while the highest is 2273 K.

In addition, Russian workers have established superplastic behaviour in Ag-Cu and Ag-Cu-Zn-Ni alloys, single and multi-phase steels including high speed cutting steel, and iron, titanium, nickel, chromium and niobium base alloys. These alloys either have compositions different from those collated in Table 3.1 or are based on new systems. (The abstracts of the publications are available in Metals Abstracts, 1970-78.) These have not been included in the Table due to the non-availability of the required data in the standard literature in English.

In this chapter the mechanical properties of structurally superplastic materials will first be considered. This will be followed by a description of microstructural studies. The chapter concludes with an examination of fracture in superplastic materials. The properties of superplastically formed alloys, on the other hand, are summarised in Chapter 6 in view of their obvious industrial relevance.

Table 3.1

Materials for which structural superplasticity (or anomalous ductility) has been reported. Only maximum m values and elongations are given although multiple source references for the different alloys are cited.

All compositions in weight percent.

Material	Maximum m	Values % elong.	Temp. range K	References
1	2	3	4	5
Bi-44.5Pb eutectic	0.42	600	293	142
Bi-43Sn eutectic	-	1950	293	12,142
Bi-31Pb-17Sn eutectic	0.45	600	293	142
Cd-17.5Zn eutectic	0.5	400	293	11
Cd-27Zn	0.5	350	293-303	11
In-34Bi eutectic	0.76	450	293	165
Pb-5Cd	0.35	-	293	166
Pb-17.4Cd eutectic	0.6	>350	298-373	167,168
Pb-30Cd	0.35	>900	298	168
Pb-19Sn	0.5	-	293-353	169
Pb-7.9Tl	0.5	400	293	170
Sn (high purity)	0.5	-	293	169
Sn-1Bi	0.48	500	295	75,94
Sn-5Bi	0.68	1000	293	76
Sn-5Cd eutectoid	0.32	350	298	171

(Table 3.1)

1	2	3	4	5
Sn-2Pb	0.5	600	293-353	169
Sn-38Pb eutectic	0.7	> 4850	293-443	70,89,90, 130,142, 145,146, 169,172-180.
Sn-2 to 6Sb	0.4	-	413-483	181
Sn-9.8Zn eutectic	0.5	570	393-453	182
Sn-32Pb-18Cd	0.55	-	298-343	183
Sn-31.1Pb-3.4Zn	0.55	500	298-343	183
Zn-(Commercial)	0.2	409	293-343	85,184,185
Zn-0.2Al	0.8	465	296	186
Zn-0.4Al	0.5	650	293	184,185, 187-191
Zn-4.9Al eutectic	0.68	300	473-633	192-195
Zn-18Al	0.6	-	295-623	196
Zn-22Al eutectoid	0.7 (1.0-ref. 209)	2900	293-573	33,42,70, 78,79,87, 104,160, 162,171, 196-212
Zn-36Al	0.5	-	295-623	196
Zn-40Al	0.65	1300	523-573	198,213-217
Zn-50Al	0.3	1000	523-573	216,217
Zn-22Al-0.1Cu	0.65	-	523	218
Zn-22Al-4Cu	0.5	1000	293-523	219,220
Zn-22Al-0.2Mn	0.5	1000	293-523	220
Zn-0.1Ni-0.04Mg	0.51	> 980	373-523	221,222
Zn-W particles	-	100	300	223
Zn-ZnO_2	-	120	300	224
α-Pu	-	218	380	225,226
β-Pu	0.33	700	453	227,228

(Table 3.1)

1	2	3	4	5
Mg (commercial)	-	80	-	229
Mg-Al eutectic	0.8	2100	623-673	141,230, 231
Mg-30.7Cd	-	250	723	167
Mg-0.5Zr	0.3	150	773	232-234
Mg-ZK60 (commercial)	0.52	1700	503-583	235-237
Mg-6Zn-0.5Zr	0.6	1000	543-583	121,158, 235-238
Al (commercial)	0.2	6000 (torsion)	650-850	239
Al-7.6Ca	0.78	850	573-873	128
Al-17Cu	0.35	600	673-797	86,240
Al-33Cu eutectic	0.9	1150	653-798	60,92,100, 105,241-244
Al-10 to 13Si eutectic (Sodium modified casting alloy LM6)	0.4	183	723-781	96
Al-6Cu-0.5Zr	0.5	2000	673-773	245-247
Al-25 to 33Cu-7 to 11Mg	0.72	>600	693-753	248
Al-12Si-4Cu	0.4	-	773	129
Al-5.6Zn-1.56Mg	0.7	500	803	249
Al-6Zn-3Mg	0.35	400	613-633	250,251
Al-10.72Zn-0.93Mg-0.42Zr	0.9	1550	823	249
Al-4.6Mg-0.75Mn-0.2Fe-0.15Si	0.4	150	548	252
Al-5.8Mg-0.37Zr-0.16Mn-0.07Cr	0.6	> 800	793	253
Al-4.5Cu-1.5Mg-0.5Mn-0.2Fe-0.1Si	0.4	200	673	254
Al-5.6Zn-2.5Mg-1.5Cu-0.2Cr-0.2Fe-0.1Si	0.4	200	673	254
7075Al alloy	-	> 200	678	255
Ag-28.1Cu eutectic	0.53	500	948	256

(Table 3.1)

1	2	3	4	5
Cu-9.5Al	0.8	-	1073	257
Cu-9.8Al	0.7	700	813-973	258
Cu-10 to 20Mg	-	250	973	167
Cu-7P	0.5	>600	683-873	93
Cu-40Zn	0.75	>525	900	82,259
Cu-48Zn	0.9	450	773-1073	138
Cu-9.5Al-4Fe (also known as CDA 619)	0.7	>800	1073	56,257, 258,260-262
Cu-10Al-3Fe	0.6	720	1073	259
Cu-10Al-4Fe	0.6	1000	1023	263
Cu-38.5Zn-3Fe	0.53	330	773-1073	129
Cu-2.8Al-1.8Si-0.4Co (also known as CDA 638)	0.5	320	773-873	264-267
Cu-38Zn-15Ni-0.2Mn	-	200	723-838	268
IN 836 (Cu base)	0.45	>300	823	129
Be (ingot source)	0.9	130	973-1273	269
Fe-0.1C	0.9	-	1000	270
Fe-0.4C	0.6	-	1000	270
Fe-0.8C	0.35	100	923-1033	271
Fe-1.3C	0.4	500	923-1173	272
Fe-1.6C	0.45	500	923-1173	272
Fe-1.9C	0.5	500	923-1173	272
Fe-0.12C-1.97Si	0.26	150	1073-1223	101
Fe-0.14C-1.93Mn	0.60	-	1000-1073	270
Fe-0.3C-2.0Al	0.37	-	1313	273
Fe-0.44C-2.0Mn	0.8	-	1000-1073	270
Fe-0.01C-6.4Ni-0.35Nb	0.40	-	973-1073	274
Fe-0.13C-1.11Mn-0.11V	0.45	310	973-1173	101
Fe-0.2C-3.1Ni-0.29Nb	0.56	-	973-1073	274
Fe-0.34C-0.47Mn-2.0Al	0.48	372	1173-1223	101

(Table 3.1)

1	2	3	4	5
Fe-0.34C-0.47Mn-2.0Al	0.48	372	1173-1223	101
Fe-0.42C-1.87Mn-0.24Si (also known as AISI 1340)	0.65	460	1000	270
Fe-0.91C-0.45Mn-0.12Si	0.42	142	989-1190	151
Fe-0.07C-0.91Mn-0.5P-0.1V	0.31	169	1073-1223	101
Fe-0.14C-0.72Mn-0.15Si-0.14V	0.43	242	1073-1223	101
Fe-0.14C-1.16Mn-0.5P-0.11V	0.57	270	1073-1223	101
Fe-0.16C-1.54Mn-1.98P-0.13V	0.55	376	1173	101
Fe-0.18C-1.54Mn-0.9P-0.11V	0.55	320	1173	101
C-Mn steels (containing small additions of V,Nb,Al and Ti)	>0.7	184	1073-1273	275
Fe-0.03 to 0.1C-3.9Ni-3Mo-1.58Ti	0.67	820	1123-1273	102,276
Fe-0.03 to 0.1C-3.9Ni-3Mo-1.58Ti-60 ppm B	0.37	634	1123-1273	276
Fe-0.14C-0.72Mn-0.15Si-0.14V	0.43	242	1073-1223	101
Fe-26Cr-6.5Ni (also known as IN 744)	0.62	>1000	973-1293	103,134, 277-279
Fe-50Cu	0.5	-	973-1073	279
Ni (commercial)	0.38	225	1073-1093	280,281
Nichrome	0.41	190	1273	281
Ni-34.9Cr-26.2Fe-0.58Ti	0.5	>1000	1068-1128	91
Ni-38Cr-14Fe-1.75Ti-1Al	0.5	1000	1083-1253	91
Ni-39Cr-10Fe-1.75Ti-1Al	0.5	1000	1083-1253	77,91,279, 282
Ni-15.0Co-9.5Cr-5.5Al-5.0Ti-3.0Mo (also known as IN 100)	0.50	1000	1083-1343	283
Ni-16Cr-8.3Co-3.4Ti-3.4Al-2.6W-1.78Ta-1.75Mo-0.9Nb-0.1Zr-0.17C-0.01B (also known as IN 738)	0.4	500	1073-1273	147
Powder consolidated NASA-TAZ-8A	-	>600	1255	129

(Table 3.1)

1	2	3	4	5
Powder consolidated U 700	0.42	1000	1300	129
Powder consolidated NASA-TRW-VIA	-	>300	1366	129
Powder consolidated IN 713	-	230	1366	129
Ti (commercial) (also known as RC 70)	0.8	-	1173	284
Ti-8Mn	0.95	140	853-1173	138
Ti-15Mo	0.45	450	853-1173	138
Ti-4Al-0.25O_2	0.6	-	1223-1323	284
Ti-5Al-2.5Sn	0.72	450	1173-1373	284
Ti-6Al-4V	0.85	1000	1023-1273	284-286
Ti-6Al-5Zr-4Mo-1Cu-0.25Si (also known as IMI 700)	-	300	1073	167
Ti-11Sn-5Zr-2.25Al-1Mo-0.25Si (also known as IMI 679)	-	500	1073	167
Co-10Al	0.47	450	1473	287
Cr-30Co	-	160	1473	288
Zr-2.5Nb	0.6	430	900-1100	123
Zircaloy 2 with 0.11 to 2.0 O_2	0.57	2500	1073-1323	111
Zircaloy 4 with 0.11 to 2.0 O_2	0.57	2500	1073-1323	111, 284
W-15 to 30Re	0.8	260	2273	289, 290

Maximum elongation critically depends on the specimen geometry and dimensions (Section 3.1.5). Therefore comparison between the superplastic response of different alloys (in terms of elongation to fracture) should be made with caution.

The strain-rate sensitivity index m can be determined in a number of ways (Section 2.7). No attempt has been made to identify the procedure of its evaluation in case of different systems because these methods, by and large, lead to similar results.

3.1 The Mechanical Properties

3.1.1 Characteristics of deformation

Although crystalline, superplastic materials deform with an apparent viscosity in the range of 10^3-10^8 Nms^{-1} (10^4 to 10^9 poises) (70) while hot glass has the value 10^6 Nms^{-1} (10^7poises) and a solid at least around 10^{14} Nms^{-1} (10^{15} poises) (291,292). In this sense the superplastic alloy is not a typical solid. The most spectacular observation (see Sections 1.1 and 2.6) concerning superplastic deformation has been the extreme elongations obtained, frequently involving several hundred per cent. Under isothermal and constant grain size conditions, during uniaxial testing this class of materials obeys a constitutive equation of the type (see Section 2.2)

$$\sigma_t = K\dot{\epsilon}_t^m \quad .. \quad 3.1$$

where σ_t is the applied stress, $\dot{\epsilon}_t$ is the strain rate, m (the strain-rate sensitivity index) and K are material constants which critically depend on experimental conditions like temperature (T), grain size (L), strain rate etc.

Though superplasticity is a high temperature phenomenon the change from normal to superplastic mode of deformation is gradual and there is no sudden onset temperature for superplasticity. Fig.3.1 due to Ball and Hutchison (79) presents the variation of flow stress with temperature for three grain sizes under conditions where the load was effectively independent of strain. The curves cross at about 0.4 T_m (T_m is the absolute transformation temperature). Above this temperature the stress for deformation increases with increasing grain size. This point of crossover may be regarded as the temperature at which the deformation changes gradually from normal to superplastic mode as the temperature is raised. (The converse, i.e. when a cross-over is seen superplasticity is indicated, however, is not necessarily true as similar behaviour is also encountered during creep.) Similar conclusions can be drawn if hardness is used as the criterion instead of flow stress (226).

A number of the major characteristics of superplastic deformation were considered in a general way in the previous chapter, namely,

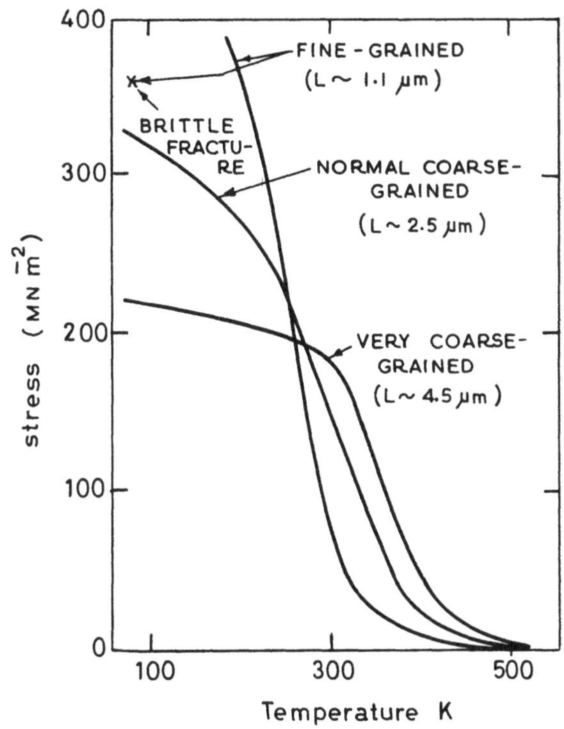

Figure 3.1 Variation of flow stress with temperature in Zn-Al eutectoid alloy of three different grain sizes (Ball and Hutchison (79))

(i) the form of the stress-strain curve (Section 2.3.1),
(ii) the variation of stress with strain-rate (Section 2.3.2),
(iii) the shape of the deforming specimen (Section 2.4), and
(iv) plastic stability and neck formation (Section 2.5).

Of these (ii) warrants further attention. Attention must also be given to the influence of temperature and grain size.

As was mentioned previously, superplastic deformation is quasi-viscous with a delay time for the onset of steady state flow. The strain necessary to attain the steady state increases with increasing strain rate and decreasing temperature. The transition from the unsteady to the steady state flow conditions, however, is gradual (93,99,209). When the stress-strain rate relation is evaluated within the steady state most superplastic materials show a sigmoidal variation of ln σ_t with ln $\dot{\epsilon}_t$. (Occasionally these sigmoidal plots are divided into three approximately linear regions of different slopes. However, on account of the smooth variation of the ln σ_t - ln $\dot{\epsilon}_t$ relationship this technique of suddenly changing the slope is perhaps unsound. The strain-rate sensitivity index, on the other hand, goes through a maximum with strain-rate (and these are shown schematically in Fig.3.2). Region II

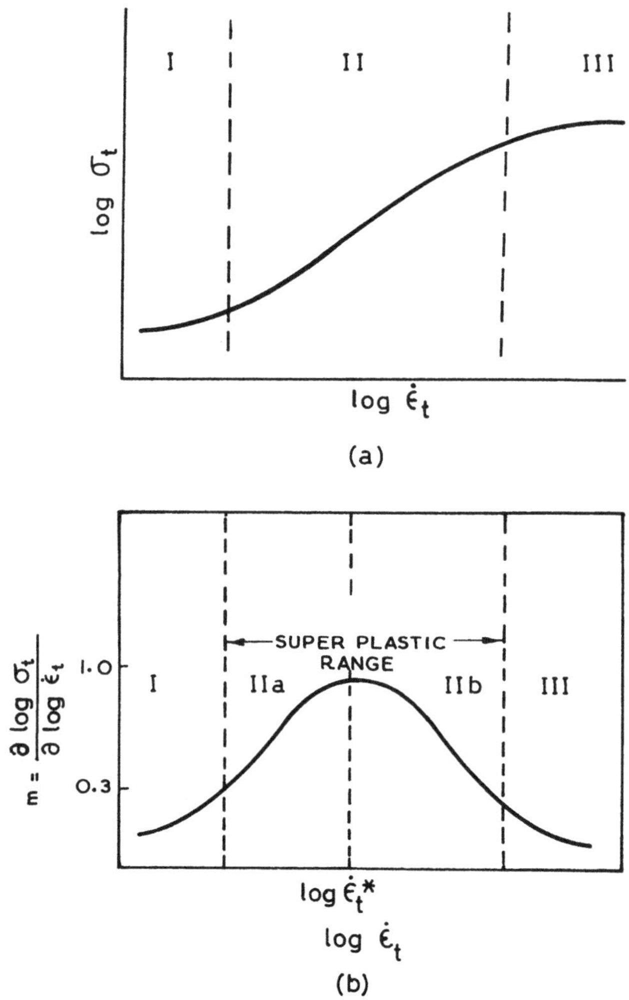

Figure 3.2 The relationship between (a) stress, σ_t, and strain rate, $\dot{\epsilon}_t$, and
(b) strain-rate sensitivity index, m and strain rate, for superplastic deformation (Schematic)

of Fig.3.2(a) with $m > 0.3$ delineates the strain-rate range over which superplasticity occurs. (Therefore, from equation 3.1 it follows that in this region strain rate increases more slowly with stress compared with regions I and III.) However, from Fig.3.2(b)it is clear that beyond a strain rate $\dot{\epsilon}_{+}^{*}$, m decreases with increasing strain rate. On account of this the region beyond a strain rate of $\dot{\epsilon}_{t}^{*}$ must be regarded as the range in which superplasticity is slowly being lost due to competition from less rate-sensitive processes. However, as in the case of decreasing temperature, the transition from superplastic deformation to a more normal mode of flow with increasing strain rate is gradual. In region IIa of Fig. 3.2(b) optimal superplastic deformation is obtained. Under these conditions the flow stress is very small and the material exhibits negligible work hardening. (It is evident that Fig. 3.2(a) is very similar to Fig.2.3. However, for a self-contained treatment inclusion of Fig.3.2(a) was found to be very useful.)

Many materials, e.g. Sn-Pb eutectic, Mg-Al eutectic, Al-Cu eutectic, Cu-9Al-4Fe, exhibit this three stage stress-strain rate behaviour. The sigmoidal $\ln \sigma_t - \ln \dot{\epsilon}_t$ curve is shifted downwards and to the right with decreasing grain size (Fig.3.3(a)) and/or increasing temperature (Fig.3.3(b)). With some alloys, e.g. Zn-22Al under certain conditions (68) and Ni-39Cr-10Fe-1.75Ti, the various stages are not well defined and the stress-strain rate curves are essentially linear over a wide range of strain rates (Fig. 3.4).

In this regard, however, it must be noted that later work (293-295) indicates three stage $\ln \sigma_t - \ln \dot{\epsilon}_t$ behaviour in case of the Zn-22Al alloy, as well. Thus it appears reasonable to conclude with Edington (80) that the absence of three stage behaviour in some superplastic alloys is perhaps due to the limited range of the experiments.

An attempt has also been made to delineate the three regions of Fig. 3.2(a) in terms of the stored energy of superplastic deformation (129). In region III the stored energy is around 12.6 J.gm atom^{-1}. This quickly falls off to zero with decreasing strain rate in region II and retains this value in region I.

Experimental data concerning Region I of Fig. 3.2(a) are sparse. Nevertheless, the decrease in m with decreasing strain rate below $\dot{\epsilon}_{t}^{*}$ indicates a lower bound strain rate for the onset of superplasticity (172). In this Region I steady state creep was preceded by long transients, the extent of which depended on the stress level (209).

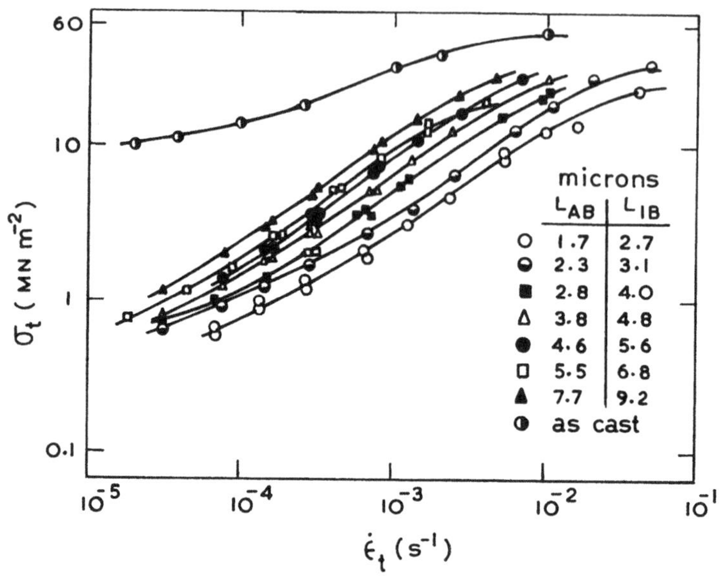

(a)

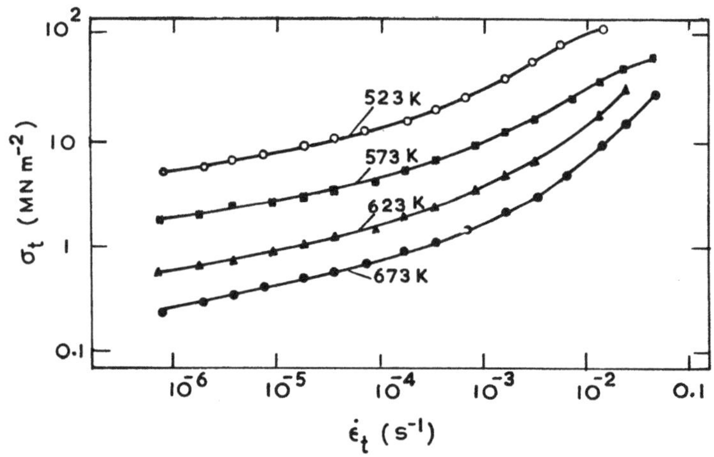

(b)

Figure 3.3(a) The stress vs. strain rate relation for various grain sizes at 793K for hot-worked and annealed Al-$CuAl_2$ eutectic alloys. The tensile strength vs. strain rate curve for as-cast material tested at the same temperature is also shown (Holt and Backofen (100)).

(b) The stress-strain rate relation for a range of temperature for the Mg-Al eutectic with grain size 2.2 μm (Lee (230))

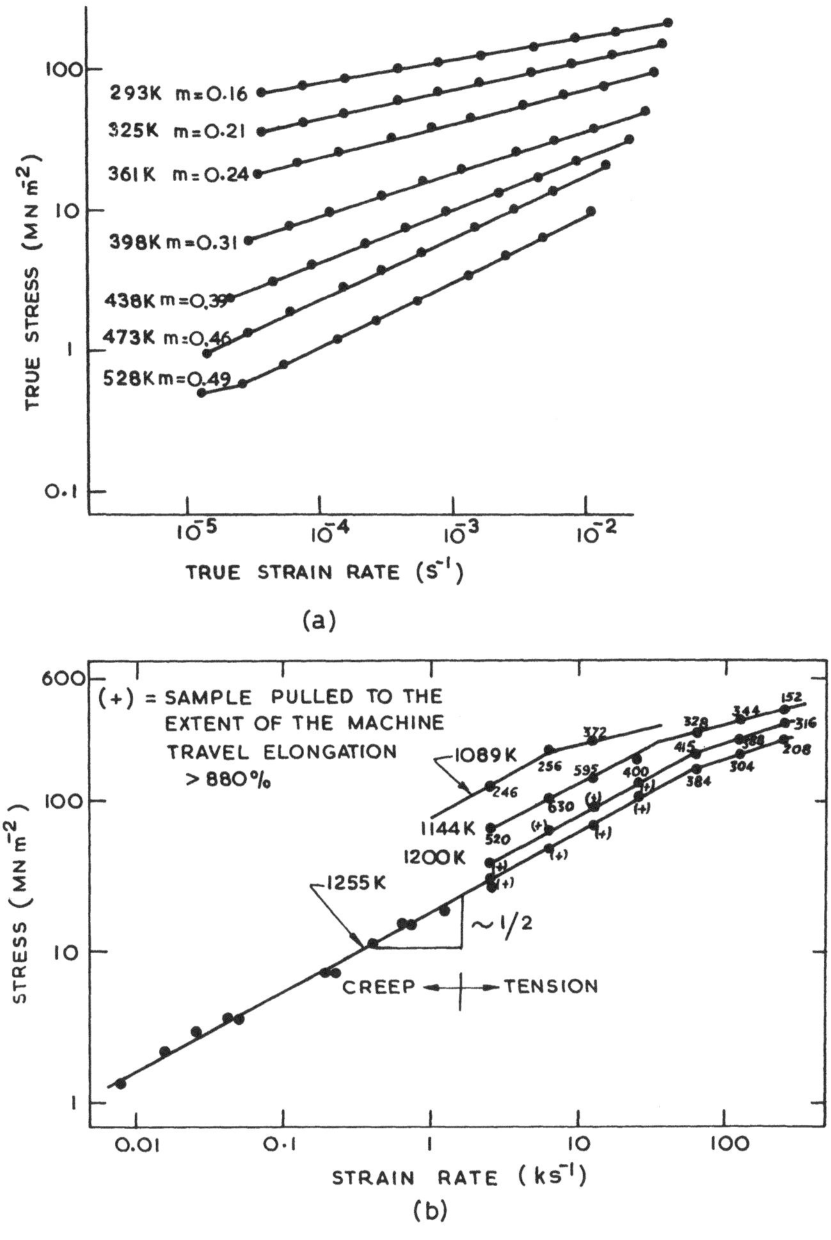

Figure 3.4 Linear variations of stress with strain rate observed in (a) the Zn-22 Al eutectoid (Nicholson (68)), and (b) the Ni-Fe-Cr system; numbers beside each point represent elongation at rupture (Hayden et al. (77))

However, there is also a school of thought that the decrease in m below a strain rate of $\dot{\epsilon}_t^*$ is merely due to the existence of an internal (296) or a threshold (62,235,296-298) stress and that at low applied stresses this gives rise to region I. Then, the equation

$$\sigma_t = \sigma_o + K\dot{\epsilon}_t^m \qquad .. \quad 3.2$$

with σ_o the internal or the threshold stress, is assumed to be the constitutive equation for superplastic flow. σ_o has been variously suggested to be constant (236,237,298), grain-size dependent (62,236) or strain-rate dependent (100). The physical arguments provided in this regard are considered in Chapter 4.

Experimentally the internal stress is evaluated by either a stress relaxation or a stress reduction test (Section 2.9.2). The limitations of the former method have already been discussed (Section 2.7.5). Similar criticisms have been levelled against the latter procedure by Geckinli and Barrett (145). The threshold stress, on the other hand, is estimated by extrapolating linear stress-strain rate plots to $\dot{\epsilon}_t = o$. This method is also unreliable because at very low strain rates the slope of these plots often change abruptly (129) and extrapolation becomes somewhat arbitrary.

Within these experimental limitations Murty (158), Lee and Niessen (222) and Geckinli and Barrett (145) have concluded that the internal stress could be significant for superplastic flow while Burton (175) and Cahoon (86) have drawn a similar conclusion with regard to the threshold stress. On the other hand, quite a few authors (60,100, 138,178,299,300) have concluded that both these concepts are yet to receive clear experimental support. Thus at present it is not clear whether it is at all necessary to use equation 3.2 in place of the simpler and certainly more popular equation 3.1.

Alternatively, in some cases Region I has been attributed to grain growth (60). This explanation is dubious because (a) grain growth during superplastic flow is rather small and (b) often region I persists over more than two decades of strain rate.

The above observations are based largely on results obtained from uniaxial tension tests. But the response of superplastic alloys does not vary significantly when the method of testing is changed to compression or torsion (66,87,99). However, as the compressive flow stress for a given strain rate is an increasing function of the diameter to height ratio, (d/h), of the specimens (99), the real compressive strength should be obtained by extrapolation to (d/h)=o of a σ_t vs. (d/h) plot at constant strain (Fig.3.5). The strain level was rela-

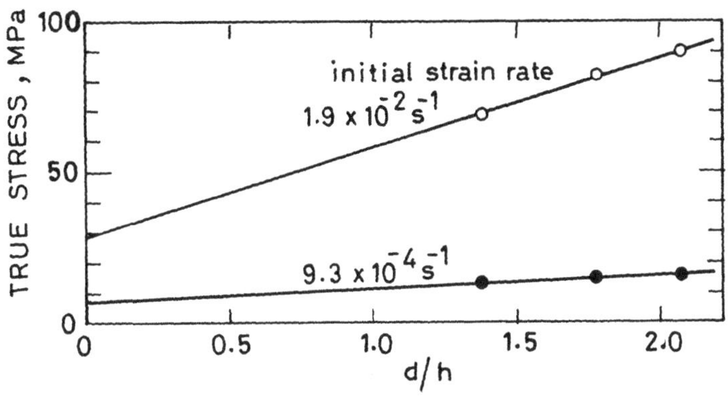

Figure 3.5 Effect of strain rate on variation of true stress with (d/h) ratio. The stress corresponding to (d/h)=0 gives the true compressive stress at the given strain rate (Padmanabhan and Davies (99).

tively unimportant so long as the amount of compression was not so great as to introduce end effects. Evidently from a series of tests at different strain rates the ln σ_t - ln $\dot{\epsilon}_t$ curve in compression can be generated. A detailed comparison of results indicates that the superplastic alloy is isotropic in its mechanical response (Fig.3.6). As in tension, the purity of raw materials used in the manufacture of the alloy and the thermo-mechanical history had negligible effect on superplastic properties if it could be ensured that the grain size was not significantly altered (see also next Section).

Moreover, unlike in the case of compression of non-superplastic materials where deformation and "barrelling" were maximum in the central portions of the specimen, during the compression of superplastic alloys the end zones underwent greater deformation and gave rise to a "double barrel" shape (132). The difference in behaviour has been traced to the high strain-rate sensitivity of the superplastic alloy.

The low cycle fatigue behaviour of a superplastic aluminium bronze has also been recently reported (301). Compared with data per-

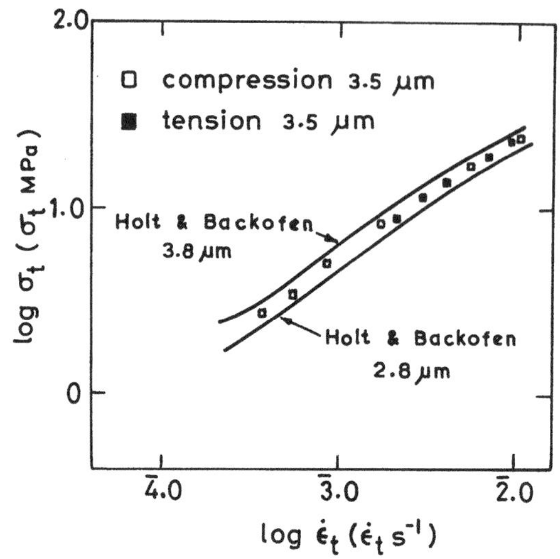

Figure 3.6 Comparison between compression and tension data at 793K for $Al-CuAl_2$ eutectic alloy (Padmanabhan and Davies (99))

taining to normal metallic materials there were essentially three differences: (a) the fatigue life increased with temperature in the superplastic range, (b) a square-shaped mechanical hysterisis loop was obtained in place of the usual lenticular configuration and (c) pronounced shape instability that made the specimens shorter and fatter was present. The differences have been explained in terms of the low flow stress, which is inversely related to the test temperature, the unique constitutive equation of flow (equation 3.1) and the exceptional ductility displayed by the superplastic alloy.

3.1.2 Necessary conditions

The existence of fine grained (< 10 μm) equiaxed polyphase structures, with the phases displaying similar ductility at the temperature of deformation, is conducive to superplasticity. The rates of deformation should be such as to give m values of 0.3 or more.

Single phase materials, e.g. cold rolled nickel (280), can be rendered superplastic by rapid testing. Duplex and multiphase structures have stable grain sizes which are resistant to grain growth and these alloye are significantly more superplastic. The type of phase or phases, their distribution, the grain-boundary condition, temperature, strain rate and grain size all influence the degree of superplasticity observed. The purity of the base metals from which alloys are prepared does not appear to affect the superplastic tendency provided the grain size is the same (173). Heat treatment does not change the functional dependence of the variables (172). There is also an empirical rule (4) that for a given base metal the lower the eutectic (or equivalent) temperature of the alloy the better the superplastic properties. Ternary eutectics appear to be more superplastic than binaries (248). If the strain-rate is set to match the velocity of diffusion processes, then most pronounced superplastic effects are observed (302).

Although under favourable conditions even non-random initial structures give rise to superplastic effects (97,144,190,287,303), depending upon the extent of directionality present in the microstructure elongation in the range of 70 to 300 percent was necessary before isotropic superplastic deformation could be induced. For a given strain, however, the tendency towards ideal behaviour increased with decreasing strain rate and/or increasing volume fraction of the second phase in the alloy (144).

When the starting material contained a weld, e.g. sheets of Ti-6Al-4V alloy (304), the behaviour of the weld during superplastic deformation was considered to be satisfactory, although this region deformed to a somewhat lesser extent.

On the other hand, superplasticity has not been observed in many alloys with continuous solubility (4). When the grain size was coarse, as in the case of β-Pu prepared from γ-Pu (227,228), or the grain boundaries were not flat and the structure was not equiaxed and uniformly fine grained no superplastic effects could be seen (102,213, 216). Annealing the material to coarsen the structure and/or to dissolve the second phase also eliminated rate sensitivity and high elongation (77,85,282). Non-deformable second phase particles, as found in dispersion hardening alloys, eliminated the effect of fine grain size and led to cavitation (121). In iron-carbon alloys undissolved carbides (151), residual massive martensite and non-equilibrium structures (271) have been shown to prevent superplasticity. In Alloy systems, e.g. the Zn-Al eutectoid, where quenching confers ultra-fine grain size, quenching from below the invariant temperature and homogenisation destroyed superplastic tendencies (198). Moreover, when the condition $0.4T_M < T < T_M$, where T is the absolute temperature of superplastic deformation and T_M is the appropriate transformation temperature (on the absolute scale) which defines the upper limit of the two phase field for the alloy, was not satisfied or when grain growth at the temperature of deformation was very rapid, significant superplasticity was absent (204, 207,251,284). Finally, when one of the phases is high melting and/or the proportion of the second phase is very small as in the Al-Fe, Al-Si and Al-Ni systems, optimal superplasticity is not observed (4,100).

Cast or cast and homogenised alloys of eutectic, eutectoid or polyphase compositions were also not superplastic although in the wrought condition they exhibited significant superplasticity. This difference has been atributed variously to the lamellar, dendritic or interlocking nature of grain boundaries in the cast alloys (100,171,201). Polarised light microscopy (169), on the other hand, has indicated that in the as cast Sn-Pb eutectic alloy the grains of the tin-rich phase were very large.

At this stage it should be noted that according to Griffiths and Hammond (138,305,306) and Likhachev et al. (239), an ultra-fine grain size is not essential for structural superplasticity. Here, "superplasticity" is seen in coarse grained materials that contain a fine scale (Stable) subgrain network. However, in these materials only the macroscopic characteristics of structural superplasticity, viz. (a) a high value of m ($\geqslant$0.2-0.3), (b) absence of work hardening, (c) considerable ductility and (d) a temperature of deformation equal to or greater than about 0.5 T_M, are satisfied. The metallographic features, however, are typical of what is seen during creep (239), hot working or superplastic deformation at strain rates greater than $\dot{\epsilon}_t^*$ in Fig.3.2(b) (307). (The microstructural characteristics of structural superplasticity are presented in Section 3.2). Therefore, it has been suggested that superplasticity encountered in large grained materials is a residual effect arising from the high values of m persisting beyond a strain rate of $\dot{\epsilon}_t^*$. Thus the <u>macroscopic</u> features described in the previous Section form a set of <u>necessary</u> but not <u>sufficient</u> conditions so far as <u>optimal</u> structural superplasticity is concerned (307).

This argument based on the magnitude of the applied strain rate (which in turn determines the microstructural features) is considered relevant because in the absence of this distinction the extreme elongation that is achieved in hot glass (or heat-softened polymers) can also be attributed to structural superplasticity. Further, the above argument involving the magnitude of the strain rate is of the essence in understanding the mechanism(s) of optimal superplastic flow because rate-insensitive flow processes become important at strain rates greater than $\dot{\epsilon}_t^*$ (Chapter 4).

Production of ultra-fine grain size

Grain growth in pure metals and solid solutions is reasonably well understood (for a brief summary see, for example, Edington (80)). Although in this class of materials "temporary superplasticity" may be induced by rapid heating and deformation at a relatively high strain rate, long term grain size stabilization is not possible. A better situation exists in case of metals of commercial purity which can be severely worked to produce fine grain sizes which are reasonably stabilised by randomly dispersed impurity particles located predominantly at grain boundaries (85,184).

When the alloy contains second phase particles the grain boundary area is eliminated where the boundary intersects the particles and this leads to a decrease in the grain boundary energy. When this decrease is more than the reduction in the overall energy of the system resulting from grain growth there will be no driving force for boundary movement. For spherical grains the maximum grain size that can be developed at a fixed temperature is then given by the approximate expression $(4R/3f)$, where R is the initial particle radius and f is the volume fraction of the second phase. This explains why structural superplasticity (characterised by a stable ultra-fine grain size) is found predominantly in alloys that contain a significant proportion of the second phase.

In fact one of the simplest methods for producing an ultra-fine grain size is to cast alloys of eutectic, eutectoid or polyphase compositions and subsequently work them by about 60 to 80 percent to obtain an intimate mixture of the phases. Evidently in this case (on account of the large volume fraction of the second phase) stable ultra-fine grains of less than 10 μm size can be obtained even when the particles are relatively coarse and well separated. This method has been extensively employed to induce superplastic behaviour in a number of systems, e.g. Pb-Cd eutectic (168), Zn-Al eutectoid (78,79,87,104,162, 196-212), Al-Cu eutectic (60,92,100,105,241-244).

Moreover, alloys of copper (56,260-262), nickel (77,91,279, 282) and titanium (284-286) as well as some alloy steels (103,134,277-279) contain a high volume fraction of the second phase (although they are of neither eutectic nor eutectoid composition) and therefore can be processed by the above procedure. In these cases, however, often a thermo-mechanical treatment becomes necessary, e.g. copper-base alloys (56,260-262).

On the other hand, from the expression for the maximum grain size at a given temperature (see above) it follows that the second phase particles should be finer when their volume fraction is smaller. However, as the temperature of superplastic deformation are usually of the order of 0.5 T_M or more considerable coarsening of the particles is likely. Thus only fine particles that coarsen very slowly should be employed. This method has been used to advantage in case of dilute alloys based on Al-Zn-Mg (251), Al-Mg-Zr (249,253,308), Al-Zn-Mg-Zr (249,253, 308), Al-Cu-Zr (245-247,309) and Zn-Ni-Mg (221,222) systems in all of which the total alloy addition is of the order of 10 wt. percent or less. As the principle of grain size control is similar for all these systems the present discussion will primarily be with regard to the Al-Cu-Zr alloys which are of considerable commercial importance (245-247).

In the ideal case particle size distribution should be bimodal (251) with the larger particles (or particles that are far apart) promoting nucleation during recrystallization and the smaller particles ensuring a small final grain size. However, in order to avoid complex thermo-mechanical treatment, in practice a single intermediate particle size is chosen (whose magnitude is system-dependent). Nucleation rate during recrystallization, on the other hand, is increased by heavy working at high strain rates, e.g. rolling, extrusion.

Although dilute aluminium-copper alloys, e.g. Al-5.6% Cu, could be severely worked and recrystallised to obtain a very fine grain size, the structure became unstable at superplastic temperatures. Likewise, in Al-Zr alloys precipitation was non-uniform and rapid grain growth occurred on account of the large zirconium-depleted zones that were present. In the Al-Cu-Zr system, however, alloys containing 2 to 6 percent copper and a zirconium level of 0.3 percent or more could be processed to produce stable, fine grain sizes. (Here the addition of zirconium to Al-Cu alloys leads to a fine scale precipitation of $CuAl_2$ while the presence of copper ensures a uniform distribution of zirconium in the solid solution.) In these alloys the width of the zirconium-depleted zones was only of the order of the recrystallized grain size and in any case these regions contained $CuAl_2$ precipitates which were stable up to moderately high temperatures. Thus the rates of recovery and recrystallization in these alloys were considerably less. Even though the volume fraction of the second phase at the temperatures of superplastic deformation was often only of the order of 0.6 percent or less (245-247, 309) a very fine dispersion of metastable $ZrAl_3$ spheroids, whose coarsening rate at superplastic temperatures is extremely slow, conferred excellent grain size stability. For best results the molten alloy is poured from a temperature slightly higher than what is usually employed in case of other aluminium alloys and rapidly solidified in order to avoid segregation of zirconium and also retain as much of the metal in solid solution as possible. (This amounts to partially sacrificing the grain refining action of zirconium during solidification.) The casting is then given an intermediate heat treatment to precipitate $ZrAl_3$ spheroids uniformly and is finally hot rolled. As the alloys were extremely resistant to recrystallization, elongations of about 50 percent at superplastic strain rates, however, were necessary to obtain a fine equiaxed grain size. On the other hand, when the intermediate heat treatment (to precipitate $ZrAl_3$ spheroids) was dispensed with and the precipitates were introduced during the hot rolling stage itself the superplastic properties were inferior.

In the Al-Mg-Zr and the Al-Zn-Mg-Zr systems magnesium and zinc perhaps play a role similar to that of copper in Al-Cu-Zr alloys (247). However, in these alloys the intermediate heat treatment (to precipitate $ZrAl_3$) is eliminated but instead a solution treatment precedes and a cold rolling procedure follows hot rolling. Moreover, unlike in the case of Al-Cu-Zr alloys, when these alloys were heated to superplastic temperature a fine recrystallized grain size resulted (even when no superplastic deformation was involved). However, mere cold work, without the intermediate hot rolling, is less effective (80).

In the Al-Zn-Mg and the Zn-Ni-Mg systems, on the other hand, the pinning of the grain boundaries is respectively due to $Mg_3Zn_3Al_2$ and Ni_3Zn_{22}/Mg_2Zn_{11} precipitates (221,222,251).

In case of low carbon as well as low alloy steels by adding appropriate combinations of α and γ stabilisers and hot working the alloy stable, duplex structures that give rise to superplastic properties at optimal temperatures could be generated, e.g. aluminium addition to low carbon steels (101); molybdenum and titanium addition to iron-4% nickel (102,276).

The powder metallurgy route has advantages particularly when compositional variations can cause difficulties. For instance, in conventionally formed nickel-base superalloys segregation and banding are severe problems (283). Therefore, when conventional forming is difficult or when it is desired that the rate of deformation should be increased to technologically attractive levels, alloy powders produced from the melt are compacted and hot extruded in the two phase field to obtain a fine grain size (283). (The grain refining mechanism in this case is not clearly understood although it is guessed that it may have some connection with the distribution of the γ' ($Ni_3(Al,Ti,Nb)$) particles). However, the prepared powders must be free from oxide and interstitial contamination failing which brittle behaviour results.

Solid state reactions, such as spinodal decomposition (78) or rapid heating to a two phase field (284), have also been successfully employed to produce random, fine grain structures.

In low alloy steels, on the other hand, controlled thermal cycling leads to a fine grained microstructure (275).

Normal splat cooling (240,242) and a modified version (255) in which the molten alloy is sprayed in the form of a jet on to a chill block, can be used to produce thin ribbons ($\sim$10 μm thick) which after subsequent compaction and extrusion can be heat-treated to obtain stable, fine grain structures which deform superplastically.

To a limited extent, electro-deposition of alternate layers of lead and tin has been used to produce fine grain sizes in Sn-Pb alloys (172).

Finally, it is of academic interest to note that ultrafine

grain sizes can be produced in pure metals, e.g. Sn (169), Ni (280), by extrusion at 77 K and cold-rolling, respectively. As mentioned earlier, rapid grain growth during the deformation of these pure metals severely restricts the superplastic elongations obtained at elevated temperature.

3.1.3 Variables of deformation

Some attention was paid in Section 2.2 to the constitutive equations governing superplastic flow which are in effect mathematical statements describing the mutual dependence of the variables of the deformation process. Load, grain size and temperature are the three important independent variables while stress, strain rate, strain and strain-rate sensitivity index (m) constitute the list of dependent variables that need consideration.

As the stress levels are determined solely by the magnitude of the load applied, discussion in terms of the latter independent variable is felt to be unnecessary. Therefore, for obtaining a clear picture the dependent variables should be considered as functions of grain size and temperature. However, in view of the controversy concerning the temperature dependence of superplastic flow (33,68) this aspect has been considered separately.

At a given temperature, the flow stress of an alloy in the superplastic state is about 10 percent of the flow stress of the same alloy in the non-superplastic condition (65). In general stress is related to strain-rate through equation 3.1. The stress at a given strain-rate and temperature decreases with decreasing grain size (Fig.3.3(a)). The dependence of stress on grain-size can be described by the relation

$$\sigma_t \big|_{\dot{\varepsilon},T} \propto L^a \qquad .. \quad 3.3$$

where L is the grain size and a is a constant lying in the range 0.5 to 3.0 (0.5-0.9 (172), 0.7 (97), 0.7-1.2 (160), 0.75-2.0 (235), 0.9-1.25 (284), 1 (91,111,249,286,310), 1-3 (173), 2 (70,173), 2-3 (311)). It has also been suggested that a could be a function of strain (97) and that it increases with increasing strain rate. In the literature there are also available two interesting speculations which are worthy of note, viz. (a) the flow stress above which the superplastic mode of deformation changes to conventional high temperature deformation represents the point at which the grain size of the alloy is equal to the size of the sub-grains resulting from deformation (243) and (b) for a given grain size the flow stress decreases with increasing angle of the grain boundaries, i.e. misorientation between grains (312).

For a given strain rate and grain size the flow stress decreases with increasing temperature (Fig.3.3(b)). The bulk of the evidence is that the temperature dependence follows an exponential relationship from which an apparent activation energy can be deduced. The validity of this type of analysis, however, has been questioned (68). Further discussion on this aspect is available in the later parts of this Section.

(a) Strain-rate

In general, uniform strain and secondary creep rate increase with decreasing grain size (55,313). Under conditions of optimal superplasticity the relation

$$\dot{\epsilon}_t \Big|_{\sigma,T} \propto \frac{1}{L^b} \qquad .. \quad 3.4$$

with b having the values 1 to 5 (1 (60,111,284), 2 (91), 2-2.6 (144, 243,295,314), 1-3 (315), 4 (316), and 5 (270)) is applicable. It follows from simple dimensional analysis that since m is less than unity, b in equation 3.4 should be greater than a in equation 3.3.

When m was independent of strain-rate (291) b was equal to 2 over wide ranges of temperature and strain rate. Therefore, the spectrum of values obtained for b could be attributed to the simultaneous variation of m with strain rate and/or grain size. A similar comment is applicable in case of the variation of flow stress with grain size discussed earlier.

In general an increase in strain rate is equivalent to an increase in grain size and has a deleterious effect on superplastic properties (81,134). Evidently this does not apply to Region IIa of Fig. 3.2(b).

The strain rate increases exponentially with temperature and as in the case of stress an apparent activation energy is often obtained from the temperature dependence of strain rate (see below).

The claim (317) that high strain rate sensitivity can be explained without the help of explicit strain rate terms, however, need not be taken seriously as the analysis amounts merely to splitting the applied stress into two parts to achieve data fitting whose physical significance and relevance to superplasticity are questionable.

(b) Strain

Irrespective of whether the superplastic effect was achieved by grain refinement, high temperature or low strain rate, superplastic creep caused no increase in the low temperature yield stress above that of the annealed material (166). On the other hand, hardening was observed after normal rate-insensitive creep which increased with strain rate. Finer grained material, however, exhibited greater recovery. Under optimal conditions thermal or thermomechanical history did not influence superplastic properties so long as the independent variables were unaltered (42,70,76,90,100). During low temperature deformation, however, the strain-hardening exponent increased with increasing grain size (207). Occasionally, at high temperatures also slight strain-hardening has been reported (76,82,89). Watts and Stowell (242) and Herriot et al. (93), however, obtained a significant linear strain dependence of flow stress and the relationship

$$\sigma_t = \sigma_I + \alpha \epsilon_t \qquad .. \quad 3.5$$

with σ_I equal to the stress for zero strain at a given strain rate and α a slowly varying function of strain rate, is valid under these conditions. Time dependent grain coarsening was shown to be responsible for this type of dependence (68,242). It is, therefore, cautioned that if the effect of grain growth, which usually is macroscopically manifested by a strain dependence of stress, is neglected spurious $\ln \sigma_t$-$\ln \dot{\epsilon}_t$ relations, m values etc. will result. It is interesting to note in this regard that superplastic deformation was essentially strain independent when structural changes were absent (61,86). Nevertheless, there are some situations where strain is considered to be important on account of dynamical effects (74,87).

(c) Strain-rate sensitivity index

The influence of variables on the strain-rate sensitivity index, m, has been considered to some extent in Section 2.7 when both the various methods of measuring m and its physical significance were examined. It is valuable, however, to examine in detail the way in which m depends upon both the independent and the dependent variables of superplastic deformation.

In single crystals and bicrystals during low stress creep tests m-values of unity have been observed (318). During superplastic deformation m of the order of unity is rarely obtained (209) (even here for the same alloy under nearly identical conditions a much lower value of m has been quoted (293)). Although in some cases m has been reported to be independent of (68,90,241,259), to increase with (70,248) or to decrease with (111,275) increasing strain rate, in general it has a maximum value (42,70,100,169,173) at a critical strain rate, $\dot{\epsilon}_t^*$ (Fig. 3.2(b)).

m increases with decreasing grain size (70,169,259) and/or increasing temperature (100,230,259,271). Although in case of many alloys the maximum value of m is reached at a temperature just below the phase boundary defining the upper limit of the two phase field, m may go through a maximum with temperature in those systems in which significant grain growth is present (60,97). The temperature dependence of m is more in region II of Fig.3.2(a) compared with either region I or III (80) and for a given system the maximum value of m that can be obtained increases with temperature (216,253) and/or decreasing grain size (80). Typical results are given in Figs. 3.7(a) and (b). In case where the m-value is independent of strain rate it is found to increase monotonically with temperature (Fig.3.8).

In spite of the above observations there is a suggestion (144) that when microstructural variations are not neglected, m is independent of strain rate and that for superplastic deformation it has a unique value of 0.5. The balance of evidence (see, for example, Melton and Edington (217), Edington et al. (129)), however, indicates that a value of 0.5 for m merely reflects experimental limitations. Moreover, even when m was practically independent of strain rate within the experimental range (68), because of its temperature dependence the values of m deviated significantly from 0.5 well within the domain of optimal superplastic flow.

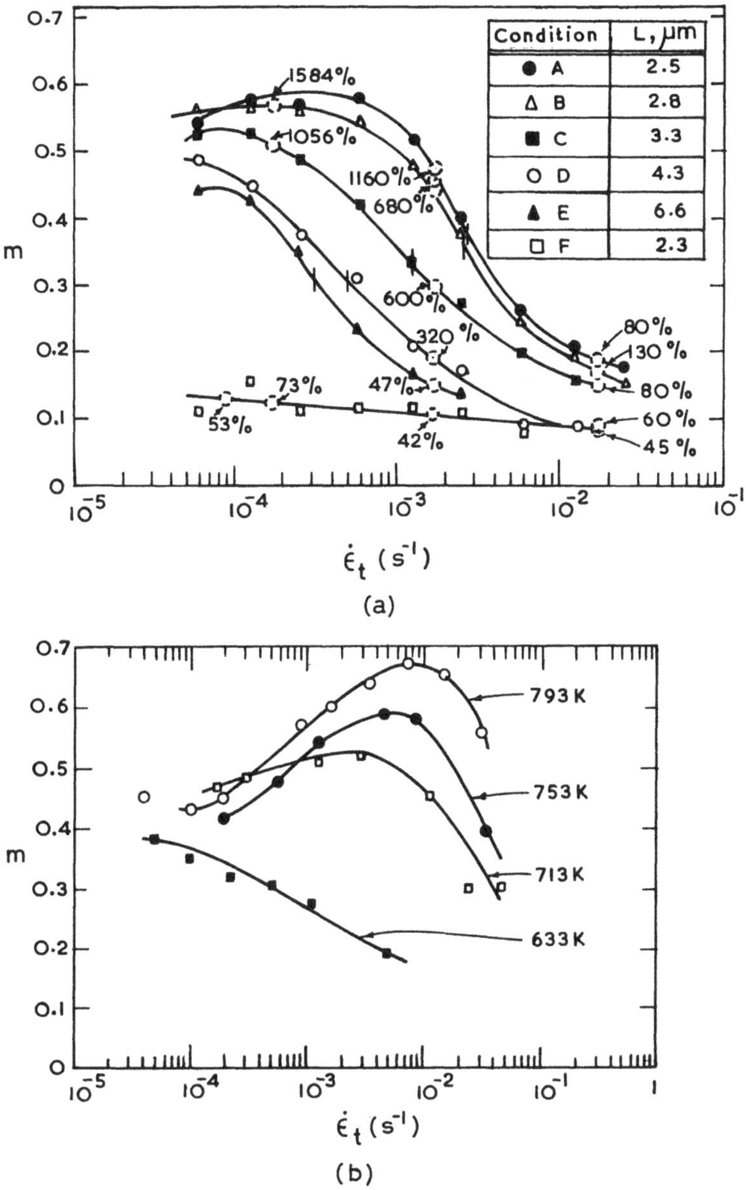

Figure 3.7 The dependence of the strain-rate sensitivity index, m, on strain rate for

(a) tin-lead eutectic specimens with different grain sizes (Avery and Backofen (70)), and

(b) aluminium-copper eutectic specimens tested at different temperatures (Holt and Backofen (100))

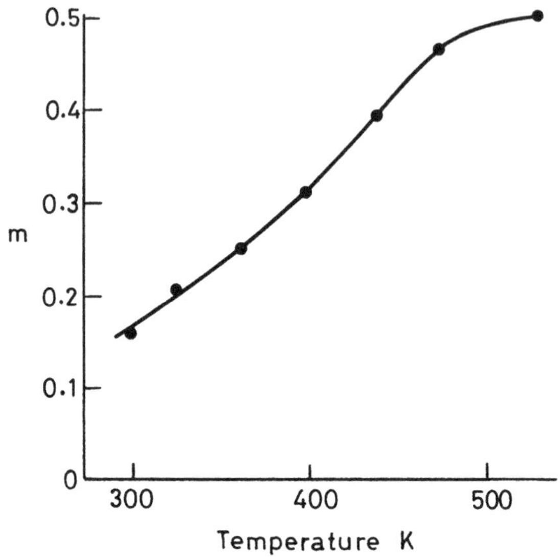

Figure 3.8 Monotonic relation between m and temperature in Zn-Al eutectoid alloy in which m is independent of strain rate at a given temperature (Nicholson (68))

Results concerning the strain dependence of the strain-rate sensitivity index are somewhat conflicting. According to Cahoon (86) and Stewart (275) m is independent of strain. But Rai and Grant (60) have concluded that m is strain-dependent at small strains and that the evaluation of m in this range will lead to spurious results. There is also some evidence that the magnitude of the applied strain rate determines the nature of variation of m with strain. At low strain rates m increases with strain (103,190) but it goes through a maximum (with strain) when the strain rate is high. At intermediate strain rates the value of m initially increases with strain to a maximum but beyond a certain point essentially strain-independent behaviour is observed (103). This difference in the strain dependence of m has been attributed to grain growth which is inversely related to strain rate (103).

The strain rate consistent with maximum m, $\dot{\epsilon}_t^*$, for a given temperature (Fig.3.2(b)) increases with decreasing grain size (97,270). Occasionally, however, the magnitude of maximum m at lower strain rates increases on increasing grain size (172). This may account for the increased stretching capacity of some superplastic alloys (42,172) on annealing. Quantitatively, the relation

$$\dot{\epsilon}_t^* \propto \frac{1}{L^c} \quad .. \quad 3.6$$

with c equal to 1 (76,172,230,258,259) or 2 (70,100) has been proposed. Suggestions exist that as $\dot{\epsilon}_t^*$ increases, the magnitude of peak m decreases (172), increases (259) or is independent of any systematic variation (264). $\dot{\epsilon}_t^*$, in general, increases with increasing temperature (100,103, 123,166,216,230,270) but there are also cases where it was independent of temperature (56,86). In Al-bronze increasing the proportion of the β-phase, adding small amounts of iron or when iron is present raising the aluminium content, increased $\dot{\epsilon}_t^*$ (257).

It is important to remember when making comparisons of data relating to m that there is definitive evidence that shows that measurements are sensitive to ambiguities arising from variations in experimental procedures and the mode of deformation (137,269,289) (see also Section 2.7). In particular microstructural coarsening can lead to erroneous results.

<u>(d) Temperature</u>

As was briefly mentioned above, the temperature dependence of both stress and strain rate appear to be exponential. Thus empirically there are two alternative expressions available to describe the temperature dependence, viz.,

$$\dot{\epsilon}_t \Big|_{\sigma, L} \propto \exp\left(-Q_\sigma/kT\right) \quad .. \quad 3.7$$

and

$$\sigma_t \Big|_{\dot{\epsilon}, L} \propto \exp\left(Q_{\dot{\epsilon}}/kT\right) \quad .. \quad 3.8$$

where the suffixes indicate the variables that are maintained constant and Q is an activation energy. However, for the above equations to be rigorously applicable to superplastic flow m should be independent of temperature. As the experimental evidence reviewed earlier suggests that the temperature dependence of m can be considerable in the region of optimal superplastic deformation both the activation energy at constant stress, Q_σ and the activation energy at constant strain rate, $Q_{\dot{\epsilon}}$, are apparent and will be different from the real activation energy for the process (319). Moreover, as m is less than unity, Q_{σ} will be greater than $Q_{\dot{\epsilon}}$ and the ratio of the latter to the former will approximately be equal to the strain-rate sensitivity index, m (69). The theoretical basis of these remarks will be discussed in Chapter 4.

In addition, when m is dependent on the test temperature equations 3.7 and 3.8 cannot be true simultaneously (319). Most of the theoretical analyses (Chapter 4) prefer to treat equation 3.7 as more exact.

Notwithstanding the above observations numerical values for the activation energy of superplastic deformation in a number of systems are available and have been collated in Table 3.2. Except for Griffiths and Hammond (138) and Heiple (269), who have estimated $Q_{\dot{\epsilon}}$, all other authors have determined Q_{σ}. Padmanabhan and Davies (69), Maulik and Padmanabhan (323) and Moles and Davies (183), however, have evaluated both Q_{σ} and $Q_{\dot{\epsilon}}$ and compared the results.

It can be seen from the Table that the activation energy for superplastic deformation is of the order of the activation energy for lattice diffusion in some systems, e.g. Zn-Al eutectic, Mg-6Zn-0.5Zr, Fe-50Cu, while it is of the order of the activation energy for grain boundary diffusion in some others, e.g. Sn-Pb eutectic, Zn-0.1Ni-0.04Mg, Ti-8Mn. Moreover, in case of most of the alloys in region III of Fig. 3.2(a) the activation energy for deformation is of the order of the activation energy for bulk diffusion (111,293,294,314,321). However, when m is high the activation energy for superplastic deformation is of the order of one half of the activation energy for volume diffusion (322), i.e. close to the activation energy for grain boundary diffusion in the alloy.

For the same system depending upon composition, stress and temperature the activation energy can be of the order of that for either lattice diffusion or grain boundary diffusion (69,322). However, when the method of evaluation was changed from tension to stress relaxation it had no effect on the magnitude of the activation energy (146).

Q_{σ} for the tin-lead eutectic (90) and the Ti-6Al-4V (286) alloys has been claimed to be independent of both stress and temperature under conditions of nearly constant m. But when m varied perceptibly with temperature and/or stress both Q_{σ} and $Q_{\dot{\epsilon}}$ decreased by about

TABLE 3.2

Reported values of the activation energy for superplastic deformation together with measured values for diffusional processes.

All compositions in weight percent.

Material	Activation energy for superplastic deformation $kJ\ mol^{-1}$	References	Activation energy for self diffusion $kJ\ mol^{-1}$	References	Activation energy for grain boundary diffusion (Smithells (320)) $kJ\ mol^{-1}$
1	2	3	4	5	6
Bi-40Cd eutectic	50.0-54.0	321	76.4(Cd) 130.2(Bi)	320 320	39.9(Cd)
Cd-5 Pb	42.0	322	101.6-105.8(Pb)	320	65.9(Pb)
Pb-5Cd	40.3	166			
Sn-5Bi	40.3-46.2	169, 322	97.9-107.5(Sn)	320	40.3(Sn)
Sn-38Pb eutectic	16.8-48.3, 32.8-50.4, 37.4-38.2, 41.2-68.0, 48.3, 57.0-84.0	69, 90, 169, 295, 314, 315, 321, 323			
Sn-32Pb-18Cd ternary eutectic	50.7-58.2 (Q_σ) 21.5-27.6 ($Q_{\dot{\epsilon}}$)	183			
Sn-31.1.1Pb-3.4Zn ternary eutectic	56.1-68.1 (Q_σ) 19.7-25.2 ($Q_{\dot{\epsilon}}$)	183	91.6-104.6 (Zn)	320	51.7-61.3 (Zn)
Zn (commercial)	90.3	85			
Zn-0.2Al	42	324	142.8 (Al)	320	
Zn-4.9Al eutectic	84-109.2	192			
Zn-22Al eutectoid	24.8-96.6, 59.2, 64.7, 74.8-121.0, 88.2-148.3, 54.0-84.0, 73.9-119.7	33, 79, 209, 211, 293, 295, 323			

(Table 3.2)

1	2	3	4	5	6
$Zn-15Al_2O_3$	58.8	325			
Zn-0.1Ni-0.04Mg	63.8	221	293.2 (Ni) 135.2-136.5 (Mg)	320 320	109.2-118.4 (Ni)
β-Pu	84	227	100 (δ-Pu)	320	
Mg-6Sn-0.5Zr	147	235	188.2-282.0(α-Zr)	111,123, 320	
Al-17Cu	163	86	197.8-202.9(Cu)	320	
Al-33Cu eutectic	78.5-193.6, 92.8-176.4, 168	69,100, 323			
Al-6Zn-3Mg	79.8	251			
Al-5.8Mg-0.37Zr	156.2	253			
Al-5.65Zn-1.56Mg- 0.41Zr	89.0	249			
Al-9.3Zn-1.03Mg-0.22Zr	74.3	249			
Al-10.72Zn-0.93Mg- 0.42Zr	74.3	249			
Cu-38Zn-15Ni-0.2Mn	143.6-147	268			
Uranium	132.3	326	168(α-U)	320	
Be (ingot source)	63.0-147.0	269			
Fe-26Cr-6Ni	220.0-280.0, 252.0-268.8	134,279	210-280(α-Fe) 268.8-306.0(γ-Fe) 307.4-309.5(Cr)	134,320 134,320 320	163.8-168.0(Fe) 193.2 (Cr)
Fe-50Cu	226.8	279			
Ni-35 to 39Cr- 10 to 26Fe alloys	134.0,151.2- 252,252	69,77, 91,279			

(Table 3.2)

1	2	3	4	5	6
Ni-15Co-9.5Cr-5.5Al-5.0Ti-3.0Mo (also known as IN100)	399.0	327			
Ti-8Mn	84	138	123.1(α-Ti) 119.7-226.8(Ti)	320 328-330	
Ti-5Al-2.5Sn	210.0-273.0	284			
Ti-6Al-4V	189.0±18.9	286			
Zr-2.5Nb	92-220	123			
Zircaloy	270.0-352.0	111			

35 to 50 percent as the stress or the strain rate was increased (69,111, 211,269,322,323). $Q_{\dot{\epsilon}}$ values, however, were considered to be more consistent (69). For a number of systems, e.g. Zn-Al eutectoid, Sn-Pb eutectic, Al-Cu eutectic, the relationship $(Q_{\dot{\epsilon}}/Q_{\sigma}) = m$ is valid (69, 183,323).

Moreover, the stress and the strain rate dependence of the apparent activation energy for superplastic deformation is significantly greater than that encountered in case of either high temperature creep or hot working (323). The linear stress dependence of Q_{σ} commonly observed during high temperature (low stress) creep (331,332) also becomes non-linear at stress levels characteristic of superplasticity. The dependence of $Q_{\dot{\epsilon}}$ on the applied strain rate, however, was linear over a limited range of strain rates (323).

Results concerning the grain size dependence of the apparent activation energy are conflicting. According to Schmidt-Whitley (243) Q_{σ} is independent of grain size. However, Watts and Stowell (333) and Maulik and Padmanabhan (323) have concluded that the apparent activation energy (according to the latter both Q_{σ} and $Q_{\dot{\epsilon}}$) could be grain size dependent, particularly when m is varying with grain size. Moreover, in a recent study Hildebrand et al. (134) have demonstrated that Q_{σ} increases linearly with the matrix grain size and that in the range of grain size employed the maximum increase was about 27 percent.

In some systems (33,123,279,294) where m, the proportion of the second phase and/or the kinetics of grain growth changed with the temperature range of superplastic deformation, Q_{σ} was temperature-dependent. (Similar studies concerning the temperature-dependence of $Q_{\dot{\epsilon}}$ are not available.)

Detailed comparison of data (69,183) has revealed that $Q_{\dot{\epsilon}}$ is either of the order of the activation energy for grain boundary diffusion in the alloy or less. Q_{σ}, on the other hand, is either considerably greater than the activation energy for grain boundary diffusion (when the lowest activation energy of the component phases is taken to be indicative of the activation energy for the alloy) or is of a similar magnitude (when the activation energy for the alloy is considered to be the mean of the values pertaining to the component phases). For the Al-17Cu alloy (86), however, Q_{σ} has been estimated to be greater than even the activation energy for lattice diffusion in the $CuAl_2$ phase.

It is relevant to note at this stage that according to Nicholson (68, 291) the activation energy deduced from either the log $\dot{\epsilon}_t$ - 1/T (=Q_σ) or the m log σ_t - 1/T ($\simeq mQ_{\dot{\epsilon}}$) plots has dubious physical significance because of the temperature dependence of (a) the strain-rate sensitivity index and (b) the activation energy itself. However, the experimental evidence presented above indicates that the temperature dependence of the activation energy, particularly under stable structural conditions, is relatively small. Thus within the narrow temperature range in which optimal superplastic flow occurs the activation energy can be looked upon as effectively temperature-independent. Correlation of Q_σ and $Q_{\dot{\epsilon}}$ with the real activation energy for the process, on the other hand, is possible if analytical procedures are available which take into account the simultaneous variation of m with temperature, stress and strain rate. Such techniques are well developed (319, 334) and are reviewed in Chapter 4.

In summary, it can be said that the activation energy values reported emphasize the important role played by diffusional processes, e.g. grain and interphase boundary diffusion, during superplastic deformation. However, as two different physical processes may require similar activation energies microstructural evidence should also be considered before conclusions are drawn regarding the mechanism(s) of flow. Nevertheless, this parameter is very useful in confirming the relevance of theoretical models.

Thus on the basis of the experimental data collated in this Section it could be concluded that the equation

$$\dot{\epsilon}_t \propto \frac{\sigma_t^{1/m}}{L^b} \exp(-Q/kT) \qquad .. \quad 3.9$$

with m in the range 0.3-0.9, b usually in the range 2-3 and Q lying between the activation energies for grain boundary and bulk diffusion, best describes superplastic flow.

3.1.4 The influence of prior treatments and compositional variations

Normally, annealing (which leads to grain growth) decreases the value of strain rate for a given m (284) but increases strength and hardness (at temperatures of superplastic deformation) (85). In Zn-Al eutectoid, however, natural ageing up to six months increased superplastic tendencies. But ageing for a longer period was detrimental (207).

Working the material prior to superplastic deformation decreases the flow stress significantly (87,177,207) but increases both m and the strain rate at which the maximum value of m is encountered (130,177, 190,248,275,335). Evidently this effect is due to the grain refinement resulting from prior work.

When the proportion of the second phase is small, as in the Zn-0.4 Al alloy, prior tensile elongation of about 25 percent was necessary to ensure relative structural stability during subsequent superplastic deformation (190).

When the initial structure was random, isotropic and uniform superplastic flow resulted. The strain ratio, R (= ratio of width to thickness strain), was nearly unity and independent of strain throughout deformation (78,193,207). But when the initial structure had a preferred orientation anisotropic superplastic behaviour was encountered. The flow stress and the strain rate sensitivity index were maximum when the tensile axis was normal to the rolling direction while the minimum value of m was observed when the two directions were parallel. The strain ratio, R, was not equal to unity in this case and it increased with the angle between the tensile and the rolling directions. However, R approached unity with increasing deformation or as a phase boundary was approached. The value of m also became independent of orientation at large strains (187,189,190,193,260,270). Occasionally, when the starting material was strongly textured initially cylindrical specimens became elliptical in cross-section following superplastic deformation (194).

In general, in Zn-Al alloys small compositional variations did not influence the superplastic behaviour (196). Likewise, the presence of a small volume fraction of relatively harder Au_4Sn particles in the Sn-Pb eutectic alloy had only an insignificant effect on superplastic properties (146). In this regard Moles and Davies (183) have concluded that when the binary alloy itself is highly superplastic, e.g. Sn-Pb eutectic, introduction of a third phase of a comparable melting point either in small (Zn) or large (Cd) quantities to form the respective ternary (alloys or) eutectics did not affect the material response during superplastic flow.

In Zn-Al alloys containing aluminium in the range 0.2 to 50wt.% structural superplasticity has been established. However, the superplastic characteristics became less pronounced on either side of the eutectoid composition (163). (Residual superplastic properties at room temperature, however, were better for alloys containing very small amounts of the higher melting Al-rich phase, e.g. Zn-0.2Al was superior to the Zn-Al eutectoid alloy at room temperature (186). In general, working prior to superplastic deformation conferred superior superplastic properties compared with the method of quenching the alloy from the single phase region. However, over a narrow range of composition centred on that of the eutectoid reaction quenching was more effective (220), perhaps due to the presence of a spinodal decomposition reaction. Although the flow stress and the temperature of maximum ductility increased with increasing aluminium content in the alloy, while the strain rate consistent with maximum elongation decreased, by suitably altering the experimental variables similar superplastic properties could be induced in alloys of widely different compositions (131).

With regard to the Zn-Al system it has also been suggested (163,216) that the maximum aluminium level that will be permitted in an alloy that has to be rendered superplastic can be increased by the adoption of better casting, heat treatment and working procedures. The as-cast microstructure, the total amount of prior work, reduction per pass in rolling, time between passes, total time and temperature of rolling and rolling speed were all seen to influence subsequent superplastic behaviour.

Addition of copper (up to one percent) or manganese to the Zn-Al eutectoid alloy led to three stage log σ_t-log $\dot{\epsilon}_t$ behaviour in the ternary alloy while in the binary alloy region I (of Fig.3.2(a)) could not be detected. Increasing the copper content beyond one percent (up to 4 percent), however, did not alter the superplastic properties. Addition of small amounts of magnesium, on the other hand, increased the flow stress. When silicon was added to the Zn-Al eutectoid, the alloy became brittle (202,220,336).

In case of a sodium-modified quasi-binary eutectic alloy containing aluminium and Mg_2Si, an excess of silicon or magnesium led to a significant decrease in the superplastic properties. Magnesium, however, had a more deleterious effect (337).

In the Al-Zn-Mg-Zr and the Al-Cu-Zr systems (245-247,309) the superplastic tendency increased with zinc and copper additions when the maximum levels of these solutes did not exceed 10 and 6 weight percent, respectively. Moreover, in zinc and aluminium base alloys the addition of solute elements that raise the temperatures of recrystallization and of superplastic deformation has been resorted to (221,245-247,253). The higher temperatures of deformation (consistent with structural stability) that are now possible ensure superior superplastic characteristics. The (residual) superplastic behaviour at room temperature, however, becomes inferior when the recrystallization temperature of the alloy is raised (221).

In aluminium bronze addition of about 4% iron made the grain size resulting from the usual thermo-mechanical treatment considerably finer and also conferred superior grain growth resistance. Therefore an extension of the strain rate and the temperature ranges of optimal superplastic deformation became possible (263).

Small amounts of manganese when added to plain carbon steels were effective in increasing the value of m (270). Similarly in Zr-2.5Nb and zircaloy (111,123) the level of impurities like oxygen significantly affected the superplastic properties and also shifted the ductility maximum to higher temperatures. This is because an increase in the oxygen content shifted the α/β phase boundary to higher temperatures and therefore a greater temperature was needed to produce a structure similar to what is encountered at lower temperatures in the oxygen-free alloys. Further, on account of the higher $\alpha-\beta$ transformation temperature a given temperature of deformation would now correspond to a lower homologous temperature ($=T/T_M$). Evidently this will lead to a decrease in the superplastic properties.

It has been known for sometime that for a given base metal a ternary eutectic alloy displays superior superplastic characteristics compared with an alloy of binary eutectic composition as a result of (a) the lower melting temperature and (b) the finer microstructure encountered in case of the former. However, in both the binary and the multi-component systems superplastic properties became inferior when the composition deviated from that of the eutectic (or equivalent) transformation (142,198). This rule, however, is not obeyed when the proportion of the second phase at the eutectic or the eutectoid composition is relatively small. For example, in eutectoid steel the proportion of the second phase is only about 12 vol. percent. Rapid grain growth and a somewhat restricted temperature range of deformation prevented significant superplastic flow in the alloy. Increasing the carbon level up to 1.9 (wt.)%(maximum Fe_3C proportion 29 vol.%) progressively increased the superplastic properties. Beyond this carbon level, however, the material could not be processed to obtain a fine-grained microstructure on account of its brittleness (272), although for a carbon content of 2.3 (wt.)%, for example the volume fraction of the second phase would be 0.35.

There is also some evidence that the superplastic behaviour becomes more pronounced when the distribution of the second phase is made finer and more uniform even though its proportion remains unaltered (111).

The nature of the second phase also appears to influence superplastic properties. For example, in steels duplex structures involving the alpha and the gamma phases give rise to superior behaviour compared with α-Fe_3C combinations. Moreover, for optimal properties it was essential that the compositions of the two phases differed significantly (to ensure microstructural stability). This could be ensured by adding suitable combinations of alpha and gamma stabilisers. The proportion of the second phase, on the other hand, could be controlled within desirable limits by selecting the test temperature properly (102).

More recently it has been suggested (274,338) that in both aluminium bronze and $\alpha-\beta$ brass best superplastic properties are encountered when the alpha: beta ratio is unity. (A change in this ratio could be effected either by the addition of small quantities of a third element, e.g. cerium in case of $\alpha-\beta$ brass, or by changing the test temperature (338).) This effect has been attributed to the presence of the greatest number of $\alpha-\beta$ (interphase) boundaries per unit volume corresponding to this particular combination of the two phases (274).

3.1.5 Elongation

Although it is a striking manifestation of the occurrence of superplasticity, the elongation of superplastic alloys was not considered in Section 3.1.3 as a dependent variable of the deformation process because (a) its reproductibility even in standard specimens is rather poor, (b) specimen geometry, composition, initial grain size, strain rate, test temperature, time- and strain-induced grain growth all affect its magnitude and it is not possible at this stage to incorporate these factors into a mathematical expression and (c) this parameter does not appear to be a property of state and hence will only be of limited use in understanding superplastic deformation.

None of the analytical or empirical equations presented in Section 2.6 predicts superplastic elongations satisfactorily although the semi-empirical equation 2.27 due to Morrison is comparatively more accurate. It was further pointed out in that Section that there is a general correlation between the elongation to fracture and the strain-rate sensitivity index, m, and often the maximum values of these two parameters coincided. When m varied in a test, (depending upon the experimental conditions) the use of its lowest, terminal or mean value gave a good correlation with the elongation data. Development of prediction equations connecting m and elongation, however, was unsuccessful due to considerable scatter of experimental results (reasons for the scatter have been enumerated in Section 2.6). Numerous authors have experimentally confirmed the validity of these points (42,70,75, 115,130,258,270,284).

For a given value of m greater elongation was obtained when $(\partial m/\partial \ln \dot{\epsilon}_t)$ was positive than when it was negative. For example, if on either side of $\dot{\epsilon}_t^*$ (Fig.3.2(b)) two strain rates of equal m were chosen, considerably greater elongation was obtained at the strain rate to the left of $\dot{\epsilon}_t^*$ (70). However, in constant crosshead speed tests maximum ductility was often encountered at an initial strain rate greater than $\dot{\epsilon}_t^*$ (103,235) because in these tests with progressive deformation the strain rate tended to that for which m is maximum ($\dot{\epsilon}_t^*$).

It is obvious from Section 2.6 that specimen dimensions affect the elongation to fracture and that the above-mentioned equation of Morrison represents at present the best formulation that accounts for this effect. Moreover, the elongation of strip specimens was inferior to that of round ones (168). The presence of surface irregularities or notches also considerably decreased the elongation to fracture (130, 339).

Elongation to fracture increased with decreasing grain size (77,127,278,287) but often it went through a maximum with strain rate (42,76,111,259). Occasionally, however, a simple reciprocal relation between elongation and strain rate is reported (60,114). The strain rate corresponding to maximum ductility, on the other hand, increased with decreasing grain size (310). The effect of an increase in temperature on elongation was similar to that of a decrease in grain size (with regard to both the magnitude of elongation and the strain rate of occurrence of its maximum value) but was more striking (103,287, 340). Maximum ductility was obtained at a temperature near (4) or about 10 K above (341) the upper limit of the two phase field. In systems prone to grain growth, however, the elongation to fracture went through a maximum with temperature well before the phase boundary was reached (60,97,281). The effect of grain growth was to decrease the elongation (42) and shift the strain rate corresponding to its maximum value to the left (127).

The temperature of maximum ductility increased with increasing grain size in some cases (134) while in others an inverse relationship was seen (111). A lowering of the strain rate of deformation, on the other hand, raised the temperature of maximum ductility (111).

In region I of Fig.3.2(a) the elongation to fracture decreased with decreasing grain size or increasing temperature (127).

In alloys prone to cavitation elongation to fracture was not reproducible even in nearly identical specimens (103) because the number and the distribution of hard, cavity-nucleating particles were not similar in every case. In these alloys the level of cavitation was a function of strain rate and often the strain rate of maximum ductility was to the left of $\dot{\epsilon}_t^*$ (56) in Fig.3.2(b). The maximum ductility depended on both the highest value of m obtainable in the system and the influence of cavitation on the time to rupture (127). Cavity interlinkage and premature failure often led to low values of elongation at fracture even when m was high (96). As the severity of cavitation decreased with increasing temperature, the maximum ductility often occurred at a temperature greater than the one for which m is maximum (267).

Occasionally superplastic elongation was affected by the testing atmosphere (111). Its magnitude also increased with the amount of prior work (85,130,184,207) but neutron irradiation had a deleterious effect (278). In a Cr-30Co alloy the highest elongation was seen in specimens displaying maximum work softening (288).

When structural anisotropy was present, in some cases maximum elongation was observed when the tensile axis was normal to the rolling direction (260,261) while in others for maximum ductility the two directions were parallel (187).

β plutonium in which the fine grain size was produced by a single temperature cycle had considerably greater ductility compared with material for which repeated thermal cycling was employed (227). This was due to extensive void formation during the latter treatment.

In some systems elongation to fracture was independent of composition so long as m was unaltered (130,284,289). In others addition of elements that increased the diffusivity or refined the grain size, e.g. phosphorus, carbon and aluminium in case of steels, iron in case of aluminium bronze, led to greater elongation (101,263). The presence of high melting or hard particles (Si in Al-Si alloys, Au_4Sn in Sn-Pb eutectic), on the other hand, decreased the ductility (146, 235).

Often elongation to fracture decreased on either side of the eutectic (75) or the eutectoid (197) composition. However, in a given system the eutectic (or equivalent) composition usually has the lowest melting point. Therefore for a fixed temperature of deformation the homologous temperature for this alloy is greater than that of other compositions. Thus the effect of composition should always be considered together with the effect of temperature (168). In general those eutectic systems in which the proportion of the second phase was larger exhibited greater elongation. This rule, however, does not apply when one of the phases is brittle at the temperature of superplastic deformation (40).

Finally there is some evidence that the elongation to fracture in microduplex alloys is maximum when the ratio of the phases is approximately unity (168,258). In some cases the morphology, the size and the distribution of the second phase were also important (111).

3.1.6 Macroscopic characteristics : an assessment

Flow of structurally superplastic materials is isotropic and involves extremely low stresses. The temperature of deformation is approximately equal to or greater than 0.4 T_M. The size of the grains in these materials is usually in the micron range (1-10 μm) and remains practically unaltered during deformation. The observation of this phenomenon in over 125 metals and alloys based on vastly different alloy systems, crystal structures, chemical compositions and temperature ranges of deformation clearly demonstrates that this is indeed a unique mode of flow.

The ln σ_t - ln $\dot{\epsilon}_t$ relationship encountered during deformation is sigmoidal. The strain-rate sensitivity index, m ($= \partial \ln \sigma_t / \partial \ln \dot{\epsilon}_t$), goes through a maximum with strain rate. In the range where m lies between about 0.3 and 1.0 enormous resistance to local deformation is seen and this results in extreme tensile elongations. Thus the flow behaviour of structurally superplastic materials closely resembles that of hot glass and heat softened polymers. A change in the method of testing from tension to compression, torsion or fatigue did not appear to change the material response.

Often maximum values of m and elongation to fracture coincided. Superplastic properties, usually measured in terms of the magnitude of the flow stress, the value of m and the elongation at fracture, are improved by increasing the temperature of deformation, decreasing the grain size and/or employing those strain rates for which m is high. The flow stress and grain size are directly related while an inverse relation is found between the strain rate and grain size. When the structure was stable work hardening was negligible.

Optimal effects are obtained in duplex or polyphase alloys in which all the phases have similar properties at the temperatures of deformation. Although pure metals and single phase alloys could be rendered superplastic, considerable grain growth restricted their superplastic tendencies. Even in alloys that are significantly superplastic grain growth, if present, led to apparent strain hardening.

The casting procedure, prior treatments (thermal, mechanical or thermo-mechanical), anisotropy of initial structure, composition and the nature (crystal structure, melting point etc.) of the phases present, their proportions, sizes and distribution, resistance to coarsening at the temperatures of deformation and small additions of other elements all influence superplastic behaviour. Rarely the testing atmosphere, e.g. presence of a controlled atmosphere or vacuum, could also affect the superplastic properties.

Alloys that are prone to cavitation do not exhibit ideal flow. Presence of inert, hard or high melting phases, a coarse grain size or a cast structure was detrimental. In general, treatments that destroyed the ultrafine grain size either partially or fully were undesirable.

Superplastic behaviour is characterised best in terms of the constitutive equation

$$\dot{\epsilon}_t \propto \frac{\sigma_t^{1/m}}{L^b} \exp(-Q/kT)$$

where m lies in the range 0.3-1.0 and b usually has values in the range 2-3. The activation energy, Q, lies between those for grain boundary and bulk diffusion. It has, however, been realised that for superplastic deformation the estimation of an activation energy using conventional Arrhenius plots is only approximate. The activation energy determined at constant strain rate, $Q_{\dot{\epsilon}}$, is not equal to that evaluated at constant stress, Q_σ, and both are different from the real activation energy for the flow process. In case of a number of systems for all practical purposes the ratio of $Q_{\dot{\epsilon}}$ to Q_σ is equal to m.

Occasionally, the existence of an internal or a threshold stress for the commencement of superplastic deformation has been suggested. Experimental evidence in this regard, however, has been a subject of controversy.

3.2 Microstructural Studies

Remarkable resemblance between the microstructures prior and subsequent to deformation has necessitated detailed investigations to identify the minor changes accompanying superplastic flow. Results of qualitative microscopy comprising surface studies, investigations of internal structure, marker experiments and in situ deformation studies are presented in Section 3.2.1. Quantitative metallographic investigations concerning the contribution of grain boundary sliding to superplastic flow and the effect of deformation on the size, shape and distribution of phases as well as the crystallographic texture are described in Section 3.2.2.

3.2.1 Qualitative metallography

(a) Surface studies

These comprise some of the simplest experiments and involve the use of optical and scanning electron microscopy of deformed specimens as well as surface replica techniques. More detailed observations regarding the behaviour of the interior of the specimens, however, are necessary before conclusions based on surface studies can be accepted as representative of bulk superplastic deformation. Within this limitation the following conclusions could be drawn:

Superplastic strain rates

(i) The most surprising observation has been the retention of an apparently equiaxed grain shape following extreme elongation (85,123,253).

(ii) Offsets along the grain boundaries in fiducial lines drawn on the polished surfaces of tensile specimens have been observed extensively (85,169,173) and attributed to the occurrence of grain boundary sliding. Evidence for interface sliding could also be established by scanning electron microscopy of deformed surfaces (241). Considerable grain rotation was also present (174,196,342). Fig.3.9, due to Alden (76), illustrates these points. Moreover, there is also some evidence that the scratch displacements and grain rotation occurred to a greater extent at interphase interfaces than at intercrystalline boundaries (343).

On account of the relative movement between grains (due to sliding) many grains were displaced as a whole above or below the original surface. Development of a rumpled or granular surface in which the size of the granules is approximately equal to the grain size is common (173,183,241). These effects became more pronounced with increasing strain and made the boundary regions well delineated, broad and distorted (100,172,196,253).

Figure 3.9 Carbon replica of a deformed Sn-5% Bi alloy which shows evidence of grain boundary offsets and grain rotation. Slip lines absent (Alden (76))

New surfaces were created at the grain boundaries as a result of grains moving in or out of the specimen surface. Irregularities at the original boundaries appeared as grooves on the new surface (Fig.3.10). These new surfaces often appeared striated, particularly following large strains at low strain rates (173,235,237). In compression the striated bands were found predominantly on boundaries approximately parallel to the stress axis (235) while in tension they appeared mostly on the transverse boundaries (235,343). Moreover, for a given axial strain the width of the striated bands obtained in compression was equal to one half of the width encountered in tension (235). In case of the Zn-22Al alloy striated band formation was maximum at the α-β boundaries and negligible at the α-α boundaries. At the β-β boundaries this tendency was two to three times less compared with that of the former type of boundaries (343). Further, the width of the striated bands increased with increasing strain or decreasing strain rate (230,343).

The formation of striations has been interpreted as evidence for the occurrence of diffusional processes accompanying grain boundary sliding (173,235,237). It has also been suggested that the results imply that below a strain rate of $\dot{\epsilon}_t^*$ (Fig.3.2(b)) sliding in combination with diffusional flow can account for the entire superplastic deformation. The dependence of the width of the striated bands on

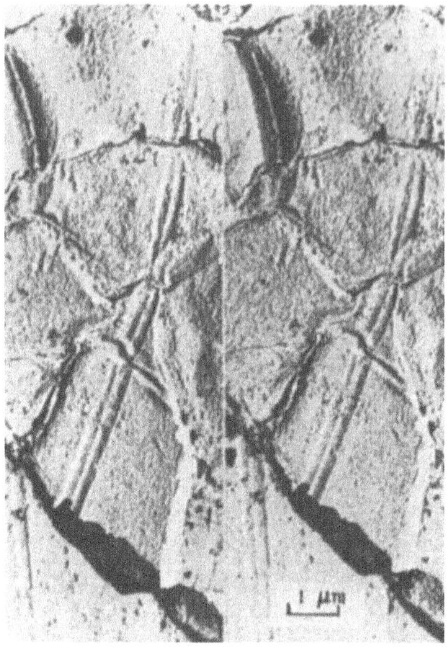

Figure 3.10 A stereo pair of surface replica from Zn-Al eutectoid specimens strained by 50% as well as in the unstrained condition. Relative movement of grains and freshly exposed regions not necessarily normal to the tensile axis are the significant features (Ball and Hutchison (79))

strain rate, on the other hand, has been interpreted as evidence for the occurrence of maximum grain boundary sliding at strain rates in the vicinity of $\dot{\epsilon}_t^*$ (343). Ball and Hutchison (79), however, have questioned this line of argument because (a) grain elongation which usually accompanies diffusion is absent during superplastic flow and (b) in their experiments the freshly exposed regions were not predominantly normal to the tensile axis. According to these authors a combination of sliding and grain rotation can explain the results. It should also be noted that two other interesting speculations regarding the origin of striations are available in the literature. According to Dunlop and Taplin (257) these are relief effects arising from the heavily gouged nature of the surfaces exposed by sliding. Lee (231), on the other hand, attributes them to the rupturing of surface oxide films across boundaries on account of grain boundary sliding.

Surface relief becomes more pronounced with deformation (Fig.

3.11). Delamination of boundaries mostly normal to the tensile axis is

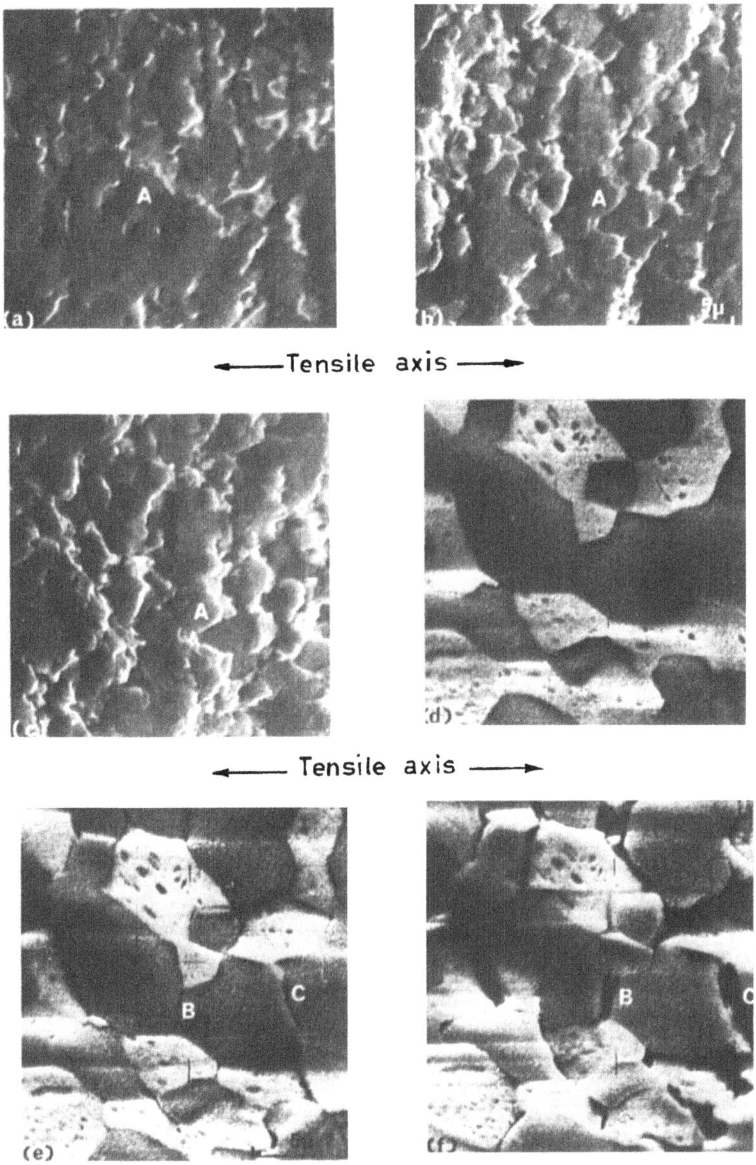

Figure 3.11 Scanning electron micrographs showing grain boundary sliding and voiding in Sn-Pb eutectic
(a) $\epsilon_t=0$, (b) $\epsilon_t=100\%$, (c) $\epsilon_t=200\%$ (A marks the same grain in (a)-(c)),
(d)-(f) correspond to strains 0, 12 and 40 percent, respectively. Voids open at B and C (Hazzledine and Newbury (61))

often seen even in highly superplastic alloys, e.g. Sn-Pb entectic - Fig.3.11. (This phenomenon is seldom observed in the interior of the specimens when the superplastic material does not contain a much harder second phase or dispersed particles.)

(iii) Within each grain fiducial lines/scratches remained straight after superplastic flow. Grain elongation was also absent. The results have been interpreted to mean that contribution from intragranular deformation to optimal superplastic flow is negligible (96,170,183). A discordant note, however, is struck by Ball and Hutchison (79) who conclude from the curvature of the scratches within the grains and the surface undulations (Fig.3.10) near the boundary offsets that intragranular deformation is important for superplastic deformation. In the light of the numerous observations made to the contrary with the aid of this and other experimental techniques (to be enumerated in due course) it is not clear whether this deduction (79) is (satistically) significant.

Non-superplastic strain rates

(iv) Experimental results regarding region I of Fig.3.2(a) are sparse. It has been suggested that both grain boundary displacements and limited grain elongation are present in this range (160,230,231). Surface scratches were considerably curved close to the sliding interfaces and freshly exposed surfaces were heavily grooved with deep gouge markings. Evidently these observations indicated a rearrangement of material near the sliding interfaces, possibly by diffusion (257).

(v) At high strain rates (Region III of Fig.3.2(a)), on the other hand, slip markings, grain elongation and a limited number of boundary offsets were seen. Freshly exposed surfaces were less grooved (160,173, 257).

(b) Internal structure

Optical metallography, replica studies and transmission electron microscopy (TEM) of interior sections of both tested and untested specimens have been extensively employed to deduce information regarding the internal structure prior and subsequent to superplastic deformation.

Retention of an approximately equiaxed grain structure following extreme superplastic elongation has been verified repeatedly using optical (197), replica (79,196) and transmission electron microscopic (78,190,211,216,230,241,272) techniques. When the initial structure was banded or consisted of elongated grains, on the other hand, superplastic flow converted the grains to an equiaxed shape (82,241,242).

In an early review (235) the evidence for grain growth during superplastic flow was considered to be equivocal. Subsequently, however, the occurrence of grain growth has been established (with the aid of the three experimental techniques mentioned above) to be a complex function of the alloy composition, test temperature, strain rate and imposed strain (see, for example, references 79,102,104,196).

Depending upon the alloy system and the experimental conditions time-dependent grain growth (127,183,242), strain-induced grain coarsening (174,196,284) or both (82,173,197,257) are reported. (These and the other results concerning grain growth presented below were established by optical microscopy.) In most cases the magnitude of both types of grain growth are relatively small (80,97,286). In one case, however, they appeared to be large (95).

In general grain growth increases with increasing strain and/or decreasing strain rate (82,190). However, there are cases where it is essentially strain-independent (191) or important only at very low strain rates (85,253,259). The occasional increase of grain growth with strain rate (68), on the other hand, has been attributed to the instability of the initial microstructure at the test temperature (127).

Grain growth is also composition dependent. In general it decreases with increasing proportion of the second phase (144,281) but in tungsten-rhenium alloys maximum grain growth was seen at a composition that gave rise to the largest elongation (289).

Coarsening of grains is considered to be the result of diffusion (131) and if this is significant the integrity of individual grains may not be preserved. Such situations, however, are rare (61).

Boundaries that were originally straight became curved following superplastic deformation. (This point has been established by both optical microscopic (103,105,174,193,207,316) and replica (196) techniques.) Subsequent annealing, however, reverted the boundaries to a flat configuration (cf. Fig.3.12(a) and (b)).

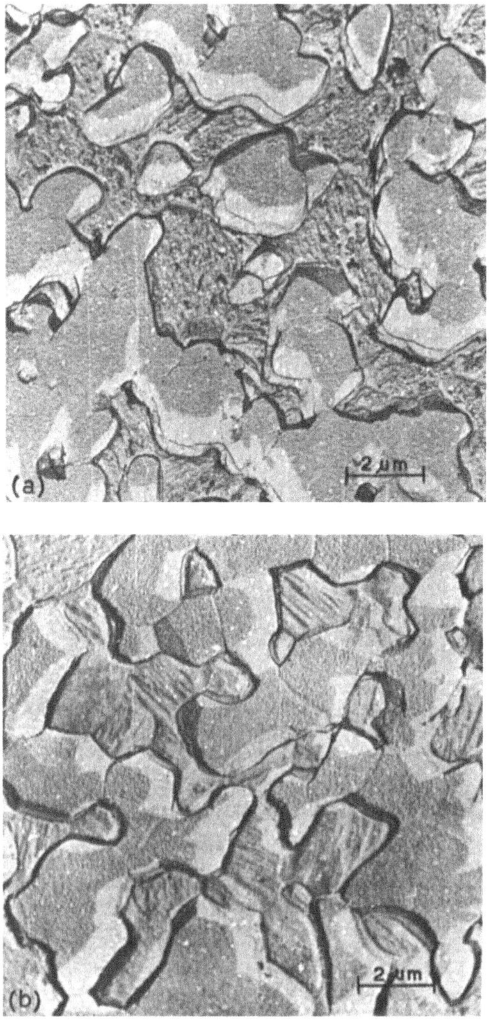

Figure 3.12 (a) Replica showing evidence for enhanced boundary curvature produced by superplastic deformation, and (b) Removal of curvature on subsequent annealing. Zn-Al eutectoid alloy (Alden and Schadler (196))

Optical metallography has further shown that in region II of Fig.3.2(a) even when the initial structure was random redistribution of the second phase (which was an increasing function of both strain and strain rate) is encountered. The interconnection of particles of the less mobile phase decreased along the tensile axis but increased in the transverse direction. Spheroidization of this latter phase was also present. Agglomeration of the second phase particles also increased with strain and/or strain rate. On enhancing the rate of deformation an increasingly discontinuous matrix phase and more irregular second phase particles resulted. The results were unaffected when the method of testing was changed from tension to compression or closed-die forging at superplastic strain rates (105,241).

With the aid of transmission electron microscopy it has been deduced that when the starting material was in the annealed condition grain and interphase boundaries were straight and triple points sharp. Each phase was a collection of individual grains (211). Following superplastic deformation at very low strain rates the boundaries remained straight (230,231). At optimal strain rates intercrystalline boundaries retained their original configuration but interphase boundaries became curved (211,230,241). Groups of grains of the same phase were redistributed so that the number of interphase interfaces became maximum, i.e. the structure became more random (211). In general the more mobile phase developed concave sides while the boundaries of the less ductile phase became convex (68,272), e.g. in Zn-Al eutectoid the (Zinc-rich) β phase developed concave boundaries; in Al-Cu eutectic the less mobile $CuAl_2$ particles got spheroidized. As a result of these minor shape changes the deformed alloy appeared to consist of mutually interpenetrating grains (61). The only effect of an increase in the test temperature, on the other hand, was to increase grain growth (204). It is also worthy of note that the final results were unaltered even when the initial microstructure was banded or consisted of elongated grains (241).

Both the development of curved boundaries and the minor changes in the shape of the grains resulting from superplastic deformation have been attributed to limited diffusional flow in the vicinity of the grain/interphase boundaries (61).

Naziri, Pearce and coworkers (104,211), however, have concluded that the post-deformation microstructure is neither uniform nor equi-axed. A number of sizes and shapes could be detected (Fig.3.13). In

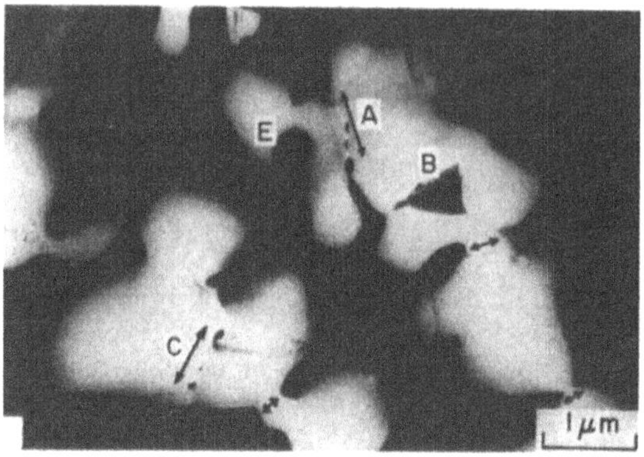

Figure 3.13 Electron micrograph (taken at 10 keV) from the gauge length of a specimen strained by 550% at 523K, and 8.4 mms^{-1}. Variations in the size and shape of grains and 'pinching off' of α grains by grains of β evident (Naziri et al. (211))

case of the Zn-Al eutectoid alloy spheroids predominantly of the Zn-rich phase were seen inside the α grains (E in Fig.3.13). 'Pinching off' of the α phase by the β phase in a random manner via the α - α boundary was common (arrowed in Fig.3.13). The results have been interpreted in terms of diffusional flow which was inseparable from grain boundary sliding and the existence of an unequal driving force across the inter-phase interfaces has been assumed.

Often deformed microstructures did not contain any disloca-tions (77,79,91,190) - Fig. 3.14. Formation of sub-grains, dislocation networks/tangles and voids was also absent (104,257,259). When the initial structure contained dislocations optimal superplastic deforma-tion caused a decrease in their density (190,201,272).

Notwithstanding these observations the suggestion that optimal superplastic deformation leads to an increase in the density of disloca-tions is occasionally encountered (216,253,254). Its validity as a general observation, however, is questionable because (a) the disloca-tions are found in very few grains which could well be the result of local stress concentrations (343), (b) in these cases initially the

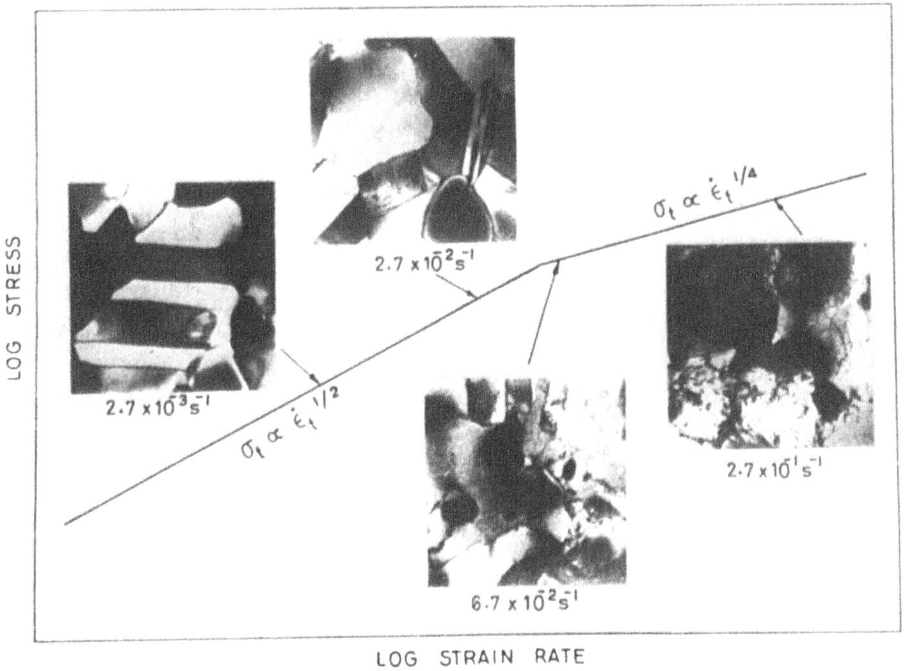

Figure 3.14 Effect of strain rate on microstructure in a Ni-Cr-Fe alloy. (a) After deformation at superplastic strain rates ($\sigma_t \propto \dot{\epsilon}_t^{0.5}$) tangles and cells are absent, and (b) following deformation at high non-superplastic strain rates ($\sigma_t \propto \dot{\epsilon}_t^{0.25}$) tangles and cells are present in good number. (Hayden and Brophy (91))

material is in a hot worked condition, and as such the presence of dislocations and/or regions containing the cast lamellar structure is not totally unexpected (163) and (c) the strain rate of deformation is in the vicinity of $\dot{\epsilon}_t^*$ in Fig.3.2(b) (216) and according to modern theories (61-63) dislocation motion commences near this value of the strain rate.

Interphase and intercrystalline cavitation, particularly the former, has been reported in a few systems, e.g. microduplex steels, copper-base alloys. This, however, is not a general feature in all the superplastic alloys. Moreover, its occurrence causes premature failure. Thus this aspect is considered in detail in Section 3.3 as part of an effort to understand the fracture of structurally superplastic alloys.

Using optical microscopy it has been shown that in region III of Fig.3.2(a) the post-deformation structure often consists of second phase particles that are aligned preferentially. When these particles had limited ductility their fragmentation was also seen. Adjacent second phase grains got agglomerated but broken pieces of the same particle did not reunite. The effects became more pronounced with increasing strain and/or strain rate (105).

With the aid of both optical (151) and transmission electron (230,231,257) microscopy the elongation of grains of all the phases has been shown to occur in region III. Transmission electron microscopy, on the other hand, has established that in this region the density of dislocations is high (79,230,231,257) - Fig.3.14(b). For fixed strain the dislocation density increased with increasing strain rate and/or grain size (77). The strain rate below which no dislocations are seen increased with temperature (79). Depending on alloy composition and test conditions dislocation arrangement could be in the form of single pile-ups, tangles, networks or cells (79,344). In region I of Fig.3.2 (a), on the other hand, to-date there is no evidence for the motion of dislocations.

Occasionally, as in the Zn-Al eutectoid alloy, a two phase structure is partially retained at temperatures clearly above the upper limit of the two phase field due to a reverse thermal hysterisis of transformation. This causes superplastic effects to persist at these temperatures (204). An increase in the test temperature (in the single phase region) and the holding time, however, decreases the hysterisis.

Observations regarding the internal structure of superplastically deformed materials have been obtained from alloys which represent a wide variety of base metals, intermetallics etc. Different methods have been employed for the production of the ultra-fine grain size. Thus the results should be regarded as general.

Nevertheless, the lack of evidence for dislocation motion need not mean that it was absent during deformation because dislocation arrays could have got relaxed or individual dislocations might have escaped to nearby boundaries or free surfaces during unloading or cooling after a test and/or thinning for microscopy. It is also conceivable that conventional methods of measuring the dislocation density may not apply to ultra-fine grained materials (294). However, (a) in situ deformation (described later) under optimal conditions also indicates that dislocation activity is negligible and (b) Lee (230) and Edington and coworkers (129) have observed a low density of dislocations even when they quenched their samples under load. Therefore, some reliance can be placed on the conclusion that dislocation activity during optimal superplastic flow is minimal.

(c) Marker experiments

In principle it is possible to introduce by suitable treatments some particles/precipitates in the superplastic alloy which serve as inert markers during deformation. Based on the changes, if any, in the size, morphology and distribution of these markers conclusions regarding the occurrence of recrystallisation, grain boundary migration and/or grain boundary sliding can be drawn. The dislocations that may get entangled with the markers, on the other hand, will provide evidence for dislocation activity. Ideally markers should be present in all the phases. In practice, however, they are often introduced only in one of the phases.

In the first experiments employing this technique (238) a Mg-6Zn-0.5Zr alloy containing many precipitates of zirconium hydride was deformed at a strain rate for which m was approximately 0.6. Formation of denuded zones at the transverse boundaries and grain rotation were seen. No evidence for massive recrystallisation, however, could be found. The denuded zones were attributed to diffusional flow whose magnitude increased with increasing strain and/or decreasing strain rate (238,345). Depending on the experimental conditions the diffusional contribution was considered to vary between 5 and 40 percent. Concomitant grain boundary sliding, however, could have influenced these estimates (see Section 3.2.2).

According to Lee (299), on the other hand, the denuded zones arise from grain boundary migration. More recently, however, it has been suggested that this interpretation is not valid for the above alloy (129). In this regard it should be noted that (a) even simple grain growth can lead to the formation of denuded zones (233) and (b) as yet no evidence exists that the concentration of zirconium at the longitudinal boundaries is greater than that found at the transverse boundaries (which should have been the case if the denuded zones were formed due to diffusional flow). But then, there is also the conclusion of Lorimer et al. (346) that conventional microproble analysis may not be applicable in case of ultra-fine grained materials. It appears, therefore, reasonable to suggest that at present conclusions regarding the origin of denuded zones cannot be considered to be unequivocal.

In the Sn-Pb eutectic alloy a stringer phase structure was produced by heat treatment (61). Superplastic deformation led to considerable redistribution of the Pb-rich phase and extensive break up of the stringers. The results were interpreted as evidence for the occurrence of extensive grain boundary sliding in the interior of the specimens. In a variation of this technique (308) pure aluminium foil was introduced between two hot rolled plates of an Al-9.3Zn-1.03Mg-0.22Zr alloy. The compound plate was then hot pressed and cold rolled to obtain tensile specimens for superplastic deformation. Oxidation of the aluminium foil provided internal markers at the interface. Electron probe microanalysis, on the other hand, indicated that the artificial introduction of the marker line did not result in the segregation of the major alloying elements, e.g. Zn, Zr. An analysis of the translation behaviour of each grain in terms of the motion of the internal marker lines suggested that during superplastic flow extensive grain boundary sliding, relative rotation of adjacent grains (necessary to accommodate sliding) and switching of (grain) neighbours were present in the interior of bulk samples. The degree of rotation was maximum at the strain rate for which m had the highest value. The magnitude of rotation, however, did not change appreciably with strain. Moreover, there is some evidence that depending on the local need to preserve the contiguity of grains the direction of rotation for each grain could change with deformation.

In a Zn-Al alloy a fine dispersion of metastable, FCC precipitates of α' was introduced in the Al-rich grains by heat treatment (68). Following superplastic deformation in the optimal range (at 448 K) the markers were unaltered with regard to their size and morphology but many Al-rich grain boundary regions devoid of precipitates were present (Fig.3.15(a)). At many of these precipitate-free zones (present at both $\alpha - \alpha$ and $\alpha - \beta$ boundaries) the shape of the patch of precipitates was similar to that of the adjacent boundary which perhaps implied that the boundary had migrated away from the edge of the preci-

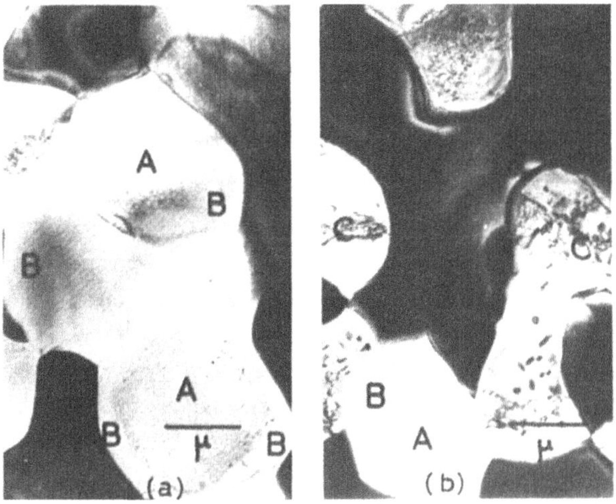

Figure 3.15 Transmission electron micrographs of deformed samples of a Zn-Al alloy containing α' (precipitate) markers in the Al-rich grains.
(a) temperature of deformation = 448K
(b) temperature of deformation = 423 K
A stands for undisturbed markers, B for precipitate-free zones and C for disturbed markers and dislocations. (Nicholson (68))

pitated area. At a lower temperature (423 K) outside the optimal range, however, appreciable coarsening and disruption of the precipitates as well as the presence of entangled dislocations could be detected (Fig.3.15(b)). The proportion of grains containing dislocations, on the other hand, increased with a decrease in the test temperature. Assuming that the behaviour of the Zn-rich grains was similar to that of the Al-rich grains the following conclusions were drawn: (a) Dislocation activity is perhaps absent during optimal superplastic flow. (b) Comparison of the dislocation arrangement seen in these experiments with that obtained from transmission electron microscopy of deformed samples (Section 3.2.1 (b)) indicates that rearrangement of dislocations perhaps takes place in case of the latter. (c) Large scale grain boundary migration, grain growth, recovery and/or recrystallisation, which would have caused the destruction of the precipitates by agglomeration with the Zn-rich grains, are absent. Thus the deformation mechanism responsible for structural superplasticity appears to leave the central part of each grain unaffected. (d) If it is assumed that the conventional methods of estimating the diffusional contribution from the size of the denuded zones are applicable in this case, then diffusion contributes about 10 percent to the total strain.

Edington and coworkers (129,216), however, have suggested that the α' precipitates may not be effective barriers for dislocations and the absence of dislocations around such precipitates need not mean that they were inactive during superplastic flow. In support of their arguments two experimental results due to Kelly and his colleagues (347, 348) have been cited, viz. (a) the deformation behaviour of single crystals of the Al-15 wt.% Zn alloy at room temperature and 77 K is consistent with the α' particles being cut by dislocations and (b) precipitation of α' leads to age softening.

This argument appears to oversimplify the implications of Nicholson's (68) findings. Fig. 3.15(a) shows that <u>the size and the morphology of the precipitates remain unchanged</u> after optimal superplastic flow. If motion of dislocations that pass through the precipitates is present, the above observation can be explained only in terms of two highly <u>unlikely events</u>, viz. (i) the precipitates are continuously recrystallising in a localised manner while the rest of the matrix does not do so (the absence of recrystallisation of the matrix is conceded by Edington and coworkers) or (ii) the precipitates are cut by such a complicated multi-slip deformation process that their shape is fortuitously restored. Moreover, outside the optimal range at a test temperature of 423 K (which is considerably higher than the temperatures employed by Kelly and coworkers) Nicholson <u>does produce</u> evidence that <u>dislocations are blocked by the α' precipitates</u> (Fig.3.15(b)). In some aluminium alloys the yield behaviour of precipitates in single crystals is significantly different from that in polycrystalline alloys of the same composition (349). It is not known whether a similar situation exists with regard to the deformation behaviour of the α' precipitates. If this is so, the preponderance of grain boundaries arising from the ultra-fine grain size and the proximity of the precipitates to the grain boundaries will only accentuate the difference. Finally, it must also be noted that the conclusion of Nicholson that the central parts of grains are not affected by optimal superplastic deformation is consistent with the numerous results obtained with the aid of a variety of experimental techniques.

Melton and Edington (216) (see also Edington et al. (129)), on the other hand, have used equilibrium zinc particles as markers in the Al-rich grains of the Zn-40 wt.%Al alloy. Comparison of the microstructure obtained after superplastic deformation with that of material simply aged for the duration of the tests revealed that enhanced coarsening and dissolution of the precipitates occurred during deformation. Evidence for the presence of dislocations was also claimed. An examination of Fig.20(b) of Edington et al. (129), however, suggests that the results indicate that the dislocation density is extremely low. Moreover, as stated in Section 3.2.1(b) the stray dislocations observed could be due to (i) the existence of local stress concentrations, (ii) the initial as extruded microstructure in which the presence of dislocations is not unlikely and/or (iii) the strain rate of deformation which is close to $\dot{\epsilon}_t^*$ (Fig.3.2(b)), in whose vicinity dislocation motion (that leads to the eventual loss of superplasticity) commences.

(d) Dynamic experiments

In situ studies that make use of transmission, scanning and high voltage electron microscopy constitute some of the most direct methods of identifying the features of superplastic deformation. The first two techniques reveal details of surface behaviour but high voltage electron microscopy (wherein specimens of about 5 μm thickness can be deformed) is likely to give results typical of bulk specimens provided the grain size is made ultra-fine.

In situ transmission electron microscopy of thin foils of Zn-Al eutectoid alloy (199) reveals that during superplastic flow grain boundary sliding and grain rotation (which aligned the boundaries in favourable orientations with respect to the stress axis) are present. But no evidence for the formation of tangles/cells or voids is available. Recrystallisation was also absent. Grain growth was negligible and the interfaces retained their integrity even when grains slid completely past their neighbours. The experimental set-up, however, did not allow the evaluation of the applied stress, strain rate etc. In any case the geometry of the specimen (particularly its thinness), the complex stress field and the observed cracking of the test sample make these observations of dubious significance to bulk superplastic flow.

Behaviour of surface grains during superplastic deformation can can be followed by scanning electron microscopy (146,257,350,351). The large depth of focus allows the details to be discerned even when the surface becomes extremely rumpled as a result of large deformation. Fig. 3.16(a) to (f) represents micrographs taken at various stages of

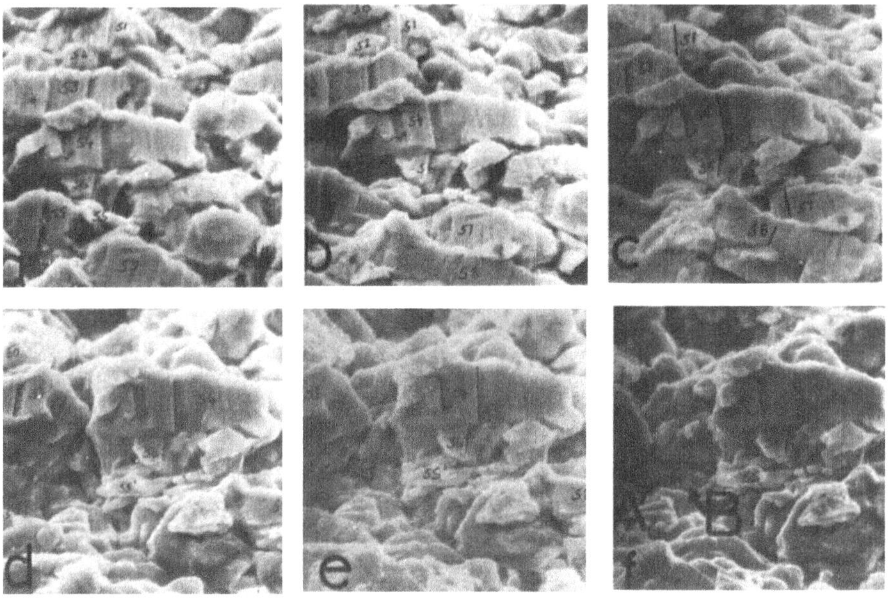

Figure 3.16 Scanning electron micrographs of a deformed sample of Sn-Pb eutectic alloy at different stages of deformation. (a)-(f) the strains involved are 120, 180, 270, 540, 610 and 800 percent, respectively. Significant grain boundary sliding is evident. (Dingley (350))

deformation of a fully aged Sn-Pb eutectic alloy. Material in the as worked condition was also studied for comparison. In the freshly worked material (where the initial structure could have been anisotropic) extensive boundary shear and movement of clumps of grains (within which there was no relative sliding) were seen. At large strains, however, grains moved individually. In the cold worked and fully aged material (closest to an isotropic structure), on the other hand, there was no evidence for the movement of <u>groups</u> of grains although sliding of individual grains was extensive. There is also a suggestion (257) that the tendency for groups of grains to move together decreased with increasing

m.

Sliding produced considerable upheavals on the specimen surface. Individuality of grains, however, was maintained although the boundaries became curved. In most cases grain growth was negligible.

Displacement due to sliding was related to the magnitude of the resolved shear stress (146,350) and boundaries oriented parallel or perpendicular to the stress axis appeared to move apart less rapidly than those of other orientations. It should be noted, however, that the latter result is complicated by the planar nature of surface observations as only the surface trace angles of the boundaries could be noted.

Grains that were originally on the surface remained at the surface throughout deformation. In most cases the direction of movement of grains was parallel to the stress axis although occasionally orthogonal displacements which led to 'grain switching' were present. Grain boundaries located at large angles to the tensile axis separated to allow the free movement of grains and also accommodate sliding. Many grains of the interior whose shape and dimensions were similar to those of surface grains appeared at the surface. These results were consistent with the occurrence of considerable grain boundary sliding at both the specimen surface and the interior (146,351). Extensive grain rotation was also present.

Under optimal conditions dislocation motion and/or recrystallisation were absent. Fiducial lines within each grain remained sharp, straight and equally spaced throughout deformation. Occasionally, however, minor changes in the shape of the grains and the presence of slip lines in a few grains that emerge from the interior are reported (351). The latter feature, however, could have been present in the starting material or arisen from local stress concentrations (both possibilities have not been considered), particularly because in the same experiments the central parts of the surface grains underwent no deformation. Nevertheless, because of the above observations a speculation does exist that diffusion and/or crystallographic slip could be accommodating grain boundary sliding.

Outside the optimal range the presence of coarse deformation bands within the grains and grain elongation have been observed. Limited grain boundary sliding occurred at the surface but no grains came from the interior to fill the sharp angular voids that developed as a result of sliding. Coalescence of these voids led to premature failure. Change in the shape of grains was perceptible. Contribution from grain strain increased with the rate of deformation and depending on the experimental conditions lay between 20 and 50 percent. Close to fracture the deformation of grains was even more extensive (146).

Sliding contribution in the optimal range has been estimated by Dingley (350) to be at least 70 percent and it has further been concluded that crystallographic slip is unimportant. More recent studies (146,321) emphasize the near equality between the imposed strain rate and the average speed at which the grains move apart. This together with the constancy of grain shape during deformation is considered to imply 100 percent contribution from grain boundary sliding. Grain rotation, a concomitant of sliding in case of superplastic alloys, provides considerable freedom in preserving the integrity of grains. Karim's (296) early suggestion that surface observations need not represent the behaviour in bulk, on the other hand, has been replaced by the later view (146,350,351) that independent movement out of the specimen surface of individual grains can only be explained by assuming that bulk movements occur in units whose size is similar to that of the surface grains. Moreover, the presence of diffusion and/or dislocation processes during bulk flow should make the grains appearing at the surface (from the interior) look elongated and this has not been reported.

<u>In situ</u> high voltage electron microscopy (500 to 1000 kV) has been made use of more recently (208,211,352) to identify the features of superplastic deformation. Specimens typically of about 5 μm thickness were heated and strained inside the specimen stage. Details of deformation were recorded on a videotape recorder via a camera mounted beneath the microscope column which viewed the electron image on a horizontal transmission screen. The recording was transferred on to a cine film for detailed frame by frame analysis and for obtaining still photographs when necessary.

Fig. 3.17(a) to (f) represents a series of results obtained in case of the Zn-Al eutectoid alloy at 373 K. Under the experimental conditions m was about 0.3. The direction of the applied stress and

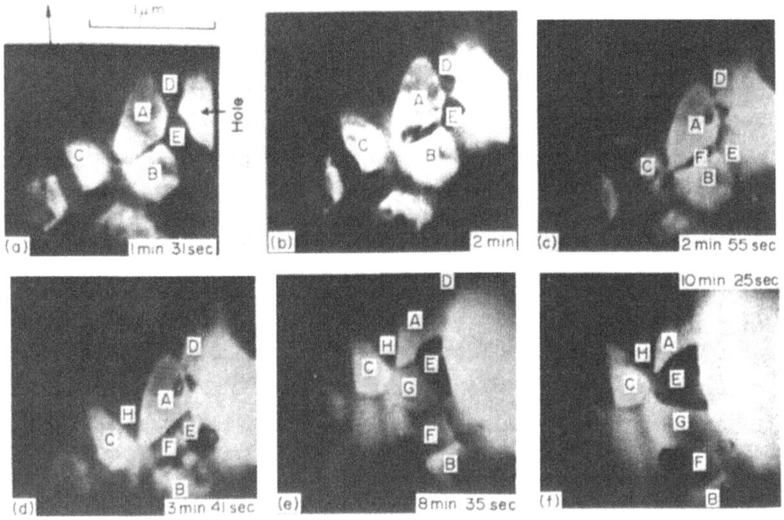

Figure 3.17 Still electron micrographs at increasing time intervals during the tensile straining of a thin foil in the microscope at 1 MeV, stress 18.5MNm^{-2}, temperature 373 K (m=0.3); grain switching, 'pinching off' of α grains by grains of β and curving of interphase boundaries present (Naziri et al. (211))

the time interval between the micrographs are shown. The following conclusions with regard to the details of optimal superplastic deformation could be drawn (208,211): (i) Extensive grain boundary sliding, particularly along interphase interfaces, occurred. (ii) Grain rotation at relatively slow rates was present. (iii) Intercrystalline boundaries remained straight during deformation. (iv) Alpha grains were 'pinched off' by the beta grains via the alpha-alpha grain boundaries (Fig. 3.17(e) and (f): two alpha grains (A and G) are 'pinched off' by two beta grains (H and E)). This process was accompanied by a gradual rounding of the interphase boundaries (convex with respect to the less ductile alpha phase grains - AE and EG in Fig. 3.17(e) and (f)) and slight changes in the configuration of the grains. (v) Random switching of grain neighbours occurred, e.g. Grains A and B in Fig.3.17 which were originally together moved away from each other in the direction of the applied stress and a beta grain came up from below and moved between the grains A and B. (vi) No dislocation motion was present in the beta grains but occasionally some dislocation activity (attributed to local stress concentrations (343)) is seen in the alpha grains. (vii) Grain growth was present. (viii) No evidence for the tearing of the foil was obtained even though there was a hole in the specimen very close to the area seen in Fig.3.17. (ix) The suggestion of Lifshitz that grain boundary sliding and diffusional creep are inseparable was experimentally verified. (x) Rapid diffusion, notably along the boundaries of the softer beta phase, was important.

Kobayashi et al. (352) have also confirmed that grain boundary sliding and diffusional flow (often manifested as boundary migration) are the two important mechanisms during optimal superplastic flow. At higher strain rates, however, dislocation motion in the vicinity of grain boundaries and triple points is seen. Initiation of cracks at triple points and their propagation along grain boundaries could also be seen.

Bricknell and Edington (353), however, have questioned the validity of results obtained by in situ high voltage electron microscopy because (a) irradiation by the electron beam could enhance the diffusivity of phases, e.g. the Al-rich phase in Zn-Al alloys, and this could change the accommodation process from crystallographic slip to diffusion; and (b) the relative importance of slip and diffusion could change when the thickness of the specimens becomes considerably more than about 5 μm. (It is, however, not clearly stated as to why there should be a difference when the thin specimen itself is a few grains thick.) According to their calculations radiation damage could occur in the Al-rich phase upto 533 K and in the Zn-rich phase up to 383 K. Experimentally, however, no damage could be seen in either phase after prolonged irradiation at 373 K, the temperature employed in the in situ studies (208,211). Even when dislocation loops were generated by irradiation at 296 K their growth stopped at about 353 K. In any case, even according to the calculations of Bricknell and Edington there should be no radiation damage at temperatures above about 533 K in case of aluminium and 763 K in case of copper. The results of Kobayashi et al. (352) concerning the Al-33 wt.% copper eutectic alloy reported above correspond to a test temperature of 793 K. Therefore they should not have been influenced by radiation damage. It is also significant that Naziri et al. (211) have concluded based on earlier creep studies under similar conditions that there was no radiation damage during their experiments.

3.2.2 Quantitative metallography

To assess the relative importance of different mechanisms to superplastic flow it is necessary to estimate their contributions quantitatively. Moreover, even a process that makes only a very small contribution could be rate controlling and therefore could determine the nature of the constitutive equation of flow. The quantitative studies have been concerned with three aspects: (a) estimating the contribution from sliding to total deformation, (b) understanding the changes in the size and shape of grains following superplastic flow and (c) analysing the variations in the intensities of texture components resulting from superplastic deformation.

(a) Grain boundary sliding

Contribution from sliding is experimentally determined by measuring the boundary offsets at the surface of deformed specimens or of those being deformed in situ using a scanning electron microscope. The marker lines are drawn either parallel (160) or perpendicular (230) to the tensile direction. Both the methods, however, are open to criticism. According to Lee (354) the use of longitudinal lines should be avoided because of the presence of normal boundary displacements during superplastic flow. Langdon and Gifkins (355), on the other hand, have suggested that transverse lines preferentially sample boundaries parallel to the stress axis and this overestimates the sliding contribution.

Within these experimental limitations a number of authors (160,308,342,350) have concluded that in the optimal range grain boundary sliding contributes about 60-70 percent to the total deformation. On either side of the optimal range of strain rate the (sliding) contribution is considerably less (76,160). Dingley and Pope (356), however, have suggested that superplastic flow is the result of 100 percent contribution from sliding.

In a detailed study Lee (230) has estimated the sliding contribution over a wide range of experimental conditions using Mg-Al eutectic alloy specimens deformed by about 17 percent. At strain rates consistent with maximum m (=0.82) sliding contributed about 64 percent to the total strain (Fig.3.18(a) and (b)). As in other systems (see above) on either side of this strain rate range the sliding contribution was considerably less (Fig.3.18(c) and (d)).

Usually the specimens are deformed by about 17-20 percent (except in one case (308) where the deformation involved was approximately 60 percent) before the boundary offsets are measured. At larger strains the accuracy of measurements is considerably less because of severe rumpling of the specimen surface. In this regard it should be recalled (Section 3.1) that when the initial structure is anisotropic elongation well in excess of 100 percent is necessary before isotropic steady state flow can be induced. Even in materials with an isotropic structure deformation is quasi-viscous and requires strains of the order of 6-10 percent for attaining the steady state (357). Therefore, it is not clear whether the numerical values obtained at relatively small strains are applicable to steady state superplastic flow involving elongations of several hundred percent.

Considerations of surface tension equilibrium require that grains intersect free surfaces at right angles. This special relationship is not seen in the bulk of the samples (296). Thus the magnitude of the contribution from sliding to total deformation may not be the same at both the surface and the interior.

During superplastic flow considerable rotation of grains as well as clumps of grains is present, particularly in the optimal range. Its magnitude could also be different at the surface and the interior. Rotation of grains, however, may not lead to specimen elongation. Thus the assumption that all boundary displacements will contribute to the total strain may not be valid (129,358).

Moreover, boundary offsets can be produced by either sliding or diffusional flow. At an early stage the offsets were attributed to diffusional creep during the occurrence of which limited sliding was

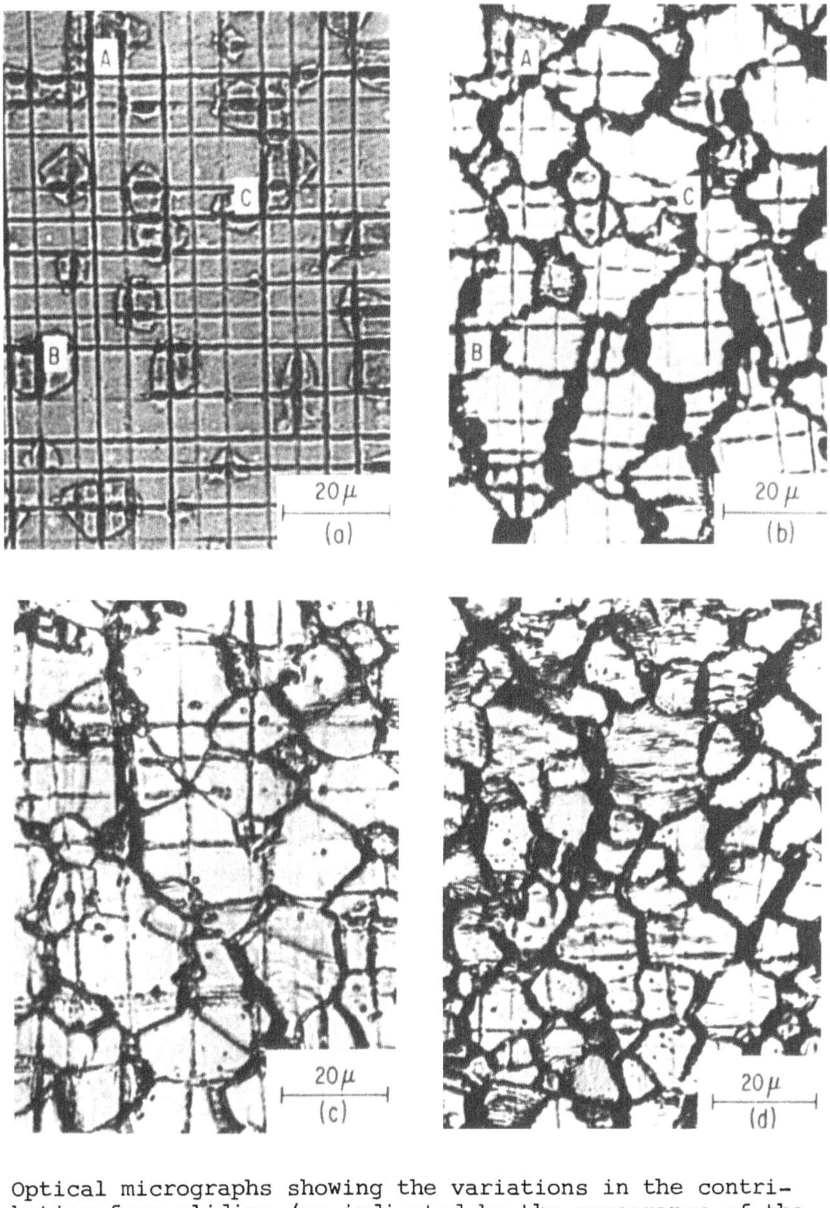

Figure 3.18 Optical micrographs showing the variations in the contribution from sliding (as indicated by the appearance of the surface) resulting from a change in strain rate. (a) unstrained material, (b) after 17% elongation at $\dot{\epsilon}_t = 6.7\times10^{-4}\ s^{-1}$ ($m \simeq 0.82$), (c) after 17.5% elongation at $\dot{\epsilon}_t = 1.7\times10^{-5}\ s^{-1}$ ($m_1 \simeq 0.29$), and (d) after 17.1% elongation at $\dot{\epsilon}_t = 3.3\times10^{-2}\ s^{-1}$ ($m \simeq 0.18$). Mg-Al eutectic alloy, tensile axis horizontal (Lee (230))

necessary to ensure geometrical compatibility (296). The more recent and popular view, however, is that a diffusion model cannot account for (a) the observed rotation of grains, (b) the non-integral (often much less than unity) values of m and (c) the decrease in m with decreasing strain rate below a value of $\dot{\epsilon}_t^*$ (Fig. 3.2(b)). Further, in most of the alloys the structure is microduplex and therefore the required diffusion fluxes would be excessively large (230,299). The above remarks, however, do not rule out limited diffusional flow particularly at low stresses, e.g. region I of Fig.3.2(a). If the latter is ignored the sliding contribution will be overestimated (359). Some error could also arise from the assumption that the contributions from sliding and diffusion are additive (129).

Even when the currently available techniques are assumed to be sound, the error in the estimates is about $\pm$ 25-30 percent (160,230, 354). It remains to be seen if the two alternative methods (355,360) proposed recently will improve the situation.

(b) Measurements of grain size and shape

Remarkable resemblance between the initial and the post-deformation microstructures was responsible for an early assumption that superplastic flow leaves the size, shape and distribution of grains unaltered. Later work, however, has shown that there are subtle, although relatively small, changes in these features. The qualitative aspects of these variations have already been covered in Section 3.2.1 (b). The results of the quantitative studies concerning the changes in the grain size and shape are summarised below. To-date no quantitative investigation is available that characterises the nature of redistribution of second phase resulting from superplastic deformation.

In single phase materials grain growth is more than twice compared with what is seen in microduplex alloys. In case of the two-phase Ni-Cr-Fe alloy (91,282) the matrix grain size was related to the second phase particle size distribution and volume fraction through a Zener-McLean type of relation. However, plots of both L vs. (2/3) (d_2/V_f) and L vs. $(2/3)(d_1/V_f)$, where L is the matrix grain size, V_f is the volume fraction of the second phase, d_1 is its diameter estimated as the arithmatic average and d_2 is the diameter calculated from the average area of the second phase particles, gave similar correlation, although in principle a superior fit should have resulted from the former. Moreover in this alloy depending on the experimental condtions grain growth was either time dependent (91) or dependent on both the time at test temperature and the imposed strain (282). When time dependent grain coarsening occurred, the increase in grain size was about 50 percent (91). Even when grain growth was determined by both time and strain, after about 800% elongation the grain size in the gauge portion was about 50 percent more than in the grip region (282). At a given strain rate the increase in grain size was proportional to the square root of the time of deformation, t. An empirical relation $(L_f-L_o)/t = K\dot{\epsilon}_t^{2/3}$, with L_f and L_o the final and initial matrix grain size and K an experimental constant, described grain growth quantitatively. Consistent with theoretical calculations regarding duplex alloys containing a second phase volume fraction of at least 18 percent in the entire experimental range near equality existed between the grain size and the mean free path of the particles.

In case of the Al-Cu eutectic alloy, on the other hand, the ratio of the grain size obtained after superplastic deformation to that produced by annealing for the duration of the test was about 1.25. The relationship $(L_f-L_o) \propto t^{0.26}$ was valid for the system (243). There is also a suggestion that when the superplastic alloy is simply annealed without the application of a stress the increase in grain size that will result will be directly proportional to the cube root of the time of annealing (310).

In a few cases the relation $(L_f^2-L_o^2) \propto t$ was valid (94,289). In the Sn-1% Bi alloy (94) deformation shifted the grain size-time (x-axis) curves upwards without changing their shape. The extent of the shift was directly related to the strain rate. The relative increases in grain size, $((L_D-L_A)/L_A)$ where L_D is the average grain size after deformation and L_A is the average grain size after annealing for the duration of the test, was linearly related to the applied (true) strain. An alternating stress system, e.g. tension-compression cycles, appeared to lead to greater grain growth compared with uniaxial tensile loading. The difference, however, was not large enough for unequivocal inference. Maximum rates of grain growth were found at strain rates for which m was large. The results have been explained in terms of the generation of excess vacancies during grain boundary sliding and by assuming that a direct relation exists between the excess vacancies generated and the applied strain rate. As the Sn-1% Bi alloy is a single phase material it is not clear whether the above relationships will hold good for microduplex alloys also.

The ability to maintain an equiaxed shape during superplastic deformation depends on the mobility of the interfaces, the extent of sliding and the deformability and the volume fractions of the phases present (129). In extruded alloys grain shape anisotropy decreased with superplastic deformation. Most of the grain shape change, however, occurred within the first 100 percent elongation (173). In isotropic materials, on the other hand, the equiaxed shape was retained to an accuracy of 15 (91) or 37 (282) percent. But the deviation from an equiaxed shape was considered to be insignificant in view of the extreme elongation of the specimens (282).

Watts and Stowell (242), however, have concluded that in case of the Al-Cu eutectic alloy superplastic deformation is not accompanied by the retention of an equiaxed structure. For the hot worked condition, although the ratio of interphase separation in the tensile direction to that in the transverse direction decreased with strain, the ratio of the grain diameters in the above two directions increased from 1.2 to 1.5 as a result of superplastic flow. In material prepared by the powder metallurgical route, on the other hand, initially both the above ratios were nearly unity but they reached a value of about 1.5 following superplastic deformation. The differences between these results and those pertaining to the Ni-Cr-Fe alloy (282) reported above have been attributed to (a) the larger volume fraction of the second phase in the former material and (b) the limited ductility of the $CuAl_2$ phase both of which make accommodation at the matrix boundaries more difficult in case of the Al-Cu eutectic alloy.

Nicholson (68), however, regards the deviation from an equiaxed shape observed by Watts and Stowell (242) as insignificant compared with what is warranted on the basis of the external change of shape of specimens. Moreover, superplastic strain could be described by an affine transformation (321). Therefore, it has been suggested that superplastic flow is an essentially random process.

It was seen in Section 3.1 that grain coarsening led to apparent strain hardening and that the increase in flow stress with strain was linear when the grain growth was time-dependent. On the other hand, when grain growth accompanying superplastic deformation was anisotropic the flow stress was directly proportional to the final grain size in the tensile direction (242).

(c) Analysis of texture

A study of texture changes accompanying superplastic deformation could provide information regarding the mechanism(s) of flow because the final texture represents the cumulative effect of the entire thermal and mechanical history (of specimens).

Using X-ray diffraction it has been shown that when the initial structure was random no preferred orientation developed as a result of superplastic flow (33,207). Texture existing prior to deformation was progressively removed by superplastic flow and following large strains the microstructure became isotropic. The rate of removal of texture, however, depended on the value of m at the given strain rate, the nature of the phase and the orientation between the tensile and the rolling directions. These results have been confirmed in case of a number of systems, e.g. both phases of the Sn-Pb eutectic alloy, the alpha phase of aluminium bronze (174,261).

In the Zn-Al eutectoid alloy when the initial structure was isotropic, the basal plane pole figures of the Zn-rich phase obtained from specimens deformed at optimal strain rates closely resembled those of the starting material (203). Deformation at higher non-superplastic strain rates, on the other hand, gave rise to sharply textured pole figures analogous to those seen in case of cold rolled zinc (Fig.3.19(a) and (b), respectively).

The above results are consistent with the occurrence of extensive grain boundary sliding and grain rotation during optimal superplastic flow. The role of crystallographic slip, however, is considered to be negligible (68).

Occasionally, when the starting structure was anisotropic, cylindrical specimens became elliptical in cross-section following superplastic deformation (194). In view of the concomitant changes in texture (187,194) this has been interpreted as evidence for the occurrence of slip during optimal superplastic flow. However, the very low intensi-

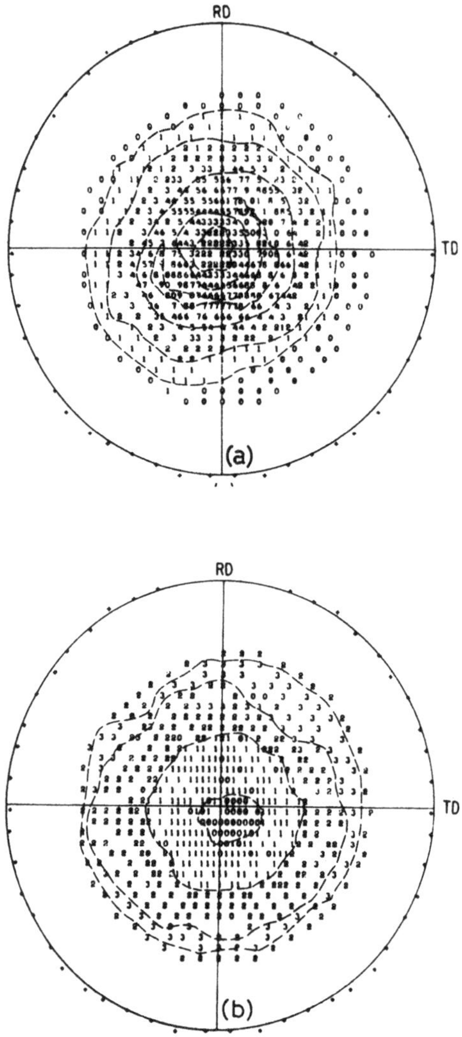

Figure 3.19 Basal plane pole figures of compressed Zn-Al eutectoid alloy specimens. (a) 81% compression at an initial (non-superplastic) strain rate of $13s^{-1}$, and (b) 85% compression at an initial (superplastic) strain rate of $1.7x10^{-3}$ s^{-1} (Lee (203))

ties of the texture peaks could be explained only with the aid of additional mechanisms like migration, recovery, recrystallisation etc., the case for each of which is highly equivocal. It is relevant to note in this regard that (a) in the above experiments (187,194) the intensities of the texture peaks decreased with superplastic deformation (b) anisotropic flow resulting from the initial anisotropic structure could result in the observed ellipticity in cross-section and (c) in polycrystalline zinc even when basal slip occurs no ellipticity in cross section is seen (132). Therefore, interpretations concerning the deformation behaviour of single crystals of HCP metals are not directly applicable to superplastic flow.

Even when the initial texture was completely destroyed by optimal superplastic flow some Russian authors (205,310,361) have attributed the observation to the occurrence of multiple slip that is 'equally matched in all directions'. The explanation is dubious because (a) in the experiments no work hardening was seen and (b) as seen above additional mechanisms like diffusion had to be invoked to explain why the presence of significant slip did not give rise to sharply textured pole figures.

In some cases (see, for example, Matsuki et al. (308)) annealing the specimens at the test temperature altered the texture while in others (214) the influence of time at the temperature of deformation was negligible.

New texture peaks were introduced when the starting structure was anisotropic (214,362). Some components, e.g. $[12\bar{1}]$ orientation of the Al-rich phase in Al-9.3Zn-1.03Mg-0.22Zr alloy (308); the peaks in the tensile and the transverse directions in case of Zn-40% Al alloy (214), were more stable than others during superplastic flow. As these components are also stable when slip occurs at low temperatures it has been suggested that crystallographic slip is important to optimal superplastic flow. The suggestion, however, cannot be regarded as unequivocal because the initial structure was anisotropic and as seen in Section 3.1.4 in materials with an anisotropic structure occasionally about 300% elongation is necessary to induce isotropic flow (187,189,190,193,260, 270,304). Thus the observed peaks could be residual or transitory (arising from compatibility requirements). Further arguments against this line of argument are available below.

The studies of texture that involve the use of pole figures (reported above) are unsatisfactory because (a) these are mostly confined to the central 70° of the stereogram on account of defocussing errors, (b) often one pole figure from one of the phases only is obtained and this prevents detailed comparison and (c) the method is unsatisfactory for understanding superplastic flow involving FCC phases (129). More recently, therefore, texture changes have been characterised in terms of the crystallite orientation distribution function (CODF) using Zn-22% Al, Zn-40% Al and Al-33% Cu superplastic alloys (210,215,244,303).

CODF is derived from normal X-ray pole figure data in the form of a series of generalised spherical harmonics. It expresses the probability that a crystallite (grain or subgrain) has a certain orientation with respect to the specimen axes. The crystallite axes are related to those of the specimen by rotation through the Euler angles θ, ϕ and ψ The CODF is normally represented by plotting the probabilities of all orientations in the Euler space. By convention constant ϕ sections are plotted after normalising the probabilities so that the units are times random. The standard deviation from random of the crystallite distribution function is called the preferred orientation parameter (POP) and this indicates the amount of texture present. The volume fraction of crystallites near a specific orientation, on the other hand, is estimated by integrating the distribution function over the corresponding region of the Euler space.

CODF's have been determined using specimens that have been subjected to maximum elongations of 200, 100 and 500%, respectively, in case of the Zn-22% Al, Zn-40% Al and Al-33% Cu alloys. The experiments covered a range of strain rates whose extremal values lay on either side of $\dot{\epsilon}_t^*$ (Fig. 3.2(b)). Fig. 3.20 represents typical ϕ = o sections corresponding to the Al-rich phase of the Zn-40% Al alloy. The CODF of the starting material has been compared with those of specimens deformed at three different strain rates. CODF of the simply annealed material and schematic diagrams showing the positions of certain ideal orientations in the ϕ = o section have also been included in the figure.

Analysis of the results revealed the following:
(i) As a result of superplastic flow level of texture in all phases decreased (due to grain rotation). (ii) The rate of texture removal was directly related to the value of the strain-rate sensitivity index because both the extent of sliding and grain rotation depended directly

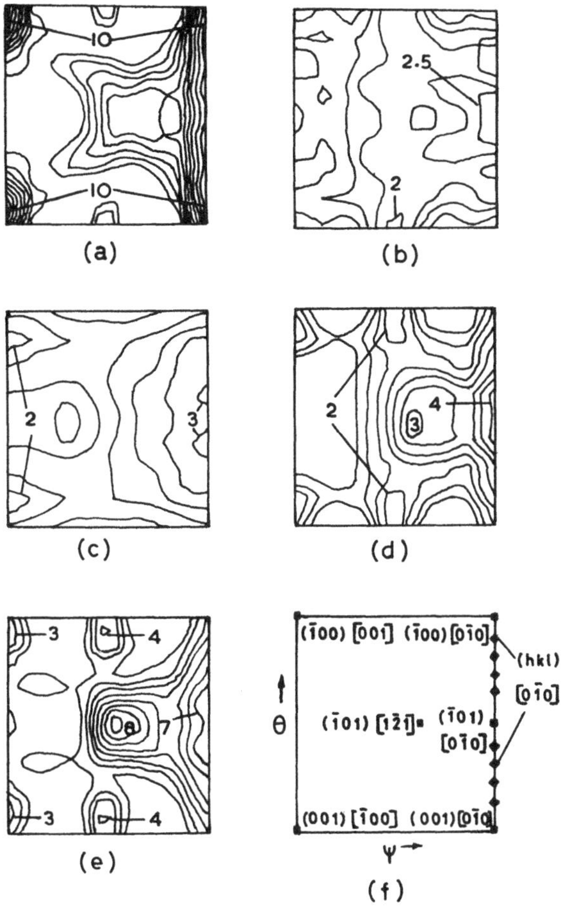

Figure 3.20 The measured $\phi=0$ sections of the crystallite orientation distribution function for the Al-rich phase (units times random) of a Zn-40 Al alloy. (a) starting material, (b) deformed 100% at $\dot{\epsilon}_t=5\times10^{-4}$ s^{-1}, (c) deformed 100% at $\dot{\epsilon}_t=5\times10^{-3}$ s^{-1}, (d) deformed 100% at $\dot{\epsilon}_t=0.05$ s^{-1}, (e) annealed for 3.6 ks at 543 K, and (f) a schematic diagram showing the position of certain ideal orientations in the $\phi=0$ section (Melton et al. (215))

on m. (iii) When the structure was isotropic superplastic flow did not give rise to new texture components. (iv) Rate of texture removal was relatively slow which implied (in accordance with the results obtained by in situ studies-Section 3.2.1(d)) that the rate of rotation of grains was small. (v) When the starting structure was anisotropic occasionally new peaks were created. (vi) In some cases annealing at the test temperature was sufficient to alter the texture. (vii) At optimal strain rates certain components of texture were more stable than others, e.g. $\{110\}$ <001> orientation of the Al-rich phase and the $\{\bar{1}010\}$ [0001] component of the Zn-rich phase in Zn-40% Al alloy. (viii) At the highest strain rates (Region III of Fig.3.2(a)) the texture results were consistent with the occurrence of multiple slip and as seen earlier (Sections 3.2.1(b) and 3.2.1(d)) electron microscopic observations are in agreement with this conclusion.

The above results indicate that grain boundary sliding and grain rotation are dominant during superplastic flow. The relative stability of some of the texture components, on the other hand, could be due to dislocation motion, anisotropic diffusional flow and/or non-random grain rotations. In view of the fact that some of the stable components seen here were also stable during low temperature dislocation motion the latter mode of deformation was considered important to optimal superplastic flow. Slip could also occur to accommodate sliding but this would not affect the texture because of the random relation between the local stress and the tensile axis. The low intensities of the texture peaks and the small changes in the orientation density, however, implied that five independent slip systems perhaps do not operate during flow under optimal conditions and that additional mechanisms, e.g. diffusion, were necessary to ensure compatibility between grains (210,244).

The above conclusions are questionable because of the following reasons: (i) When the boundaries of surface grains are either parallel or perpendicular to the stress axis they move apart less rapidly compared with boundaries of other orientations. The amount of rotation also varies from one grain to another and even in case of the same grain the sense of rotation changes during the course of a test (146). Thus the assumption of Edington and coworkers (210,215,244,303) that the grain rotations accompanying sliding are random may not be valid. (ii) When the boundary was parallel to the tensile axis no sliding occurred (243,363). Thus grain boundary sliding also could be orientation-dependent. (iii) In a given alloy the rate of removal of texture changed from one phase to another (144). (iv) The observed texture peaks could be transitional and might owe their origin to the (grain) compatibility requirements imposed by non-random grain boundary sliding and grain rotation. (v) When the starting structure was isotropic the only increase in intensity (of the texture peak) observed to-date has been that of the $\{\bar{1}010\}$ [0001] component of the Zn-rich phase in the Zn-40% Al alloy. This has, however, been attributed to anisotropic diffusion (in the Zn-rich phase) (215). (vi) The relative stability of some components in the $CuAl_2$ phase has been attributed to anisotropic diffusion (244). Thus the stability of certain components, per se, need not mean that slip is present. (vii) Often annealing at the test temperature alters the texture. Thus the changes arising from time at the test temperature should be isolated before those introduced by superplastic deformation can be identified. (viii) It has been conceded (244) that no experimental evidence has so far been produced with the aid of other techniques that dislocation motion occurs at strain rates less than $\dot{\epsilon}_t^*$. (ix) Grain switching and rotation of grains as though each grain is practically independent of its neighbours are seen during superplastic deformation. As these observations have not been reported for low temperature slip the comparison between superplastic flow and low temperature deformation is not straightforward. (x) It has not been demonstrated that a certain texture can result only from a particular mode of deformation. (xi) Hazzledine and Newbury (61) have concluded that at present it is not possible to identify the origin of the weak textures seen after superplastic deformation.

3.2.3 Assessment of metallographic and topological features

Structural superplasticity is encountered in materials having an ultra-fine grain size. The number of elements in the solid solutions, the number, chemistry and crystal structure of the component phases and/or the special properties of eutectics/eutectoids do not appear to be important in determining the superplastic characteristics. Composition influences properties because it determines the rate of grain growth.

Briefly, any satisfactory theory of superplastic deformation should take into account the following observations:

(i) The identity of individual grains is maintained during deformation, i.e. there is no massive recrystallisation.

(ii) Considerable experimental evidence has been produced with the aid of a number of techniques which clearly shows that extensive grain boundary sliding is present during optimal superplastic flow, e.g. scratch offsets, development of rough and granular appearance on the surfaces of deformed specimens, negligible grain growth, retention of an equiaxed grain shape, weakening of initial texture with superplastic deformation and a strain-rate sensitivity index of flow whose magnitude is very similar to that encountered during grain boundary sliding in bicrystals (364).

(iii) An imbalance of shear stresses acting at different boundaries (due to differences in boundary orientations, diffusivities of phases/ interfaces, atomic arrangement at intercrystalline and interphase boundaries etc.) gives rise to significant grain rotation which accommodates sliding by aligning the boundaries favourably with respect to the stress axis. In this manner additional freedom for the occurrence of grain boundary sliding is made available.

(iv) During flow under optimal conditions individual grains behave as though they are practically independent of their neighbours, subject to the constraint that their relative movements should not lead to void formation. Therefore grain boundaries that are oriented at large angles separate to allow free movement of grains and also permit new grains to come between those initially together (grain switching).

(v) Grain shape remains approximately equiaxed during deformation although grain elongation which is small in comparison with external shape changes may occur.

Although this observation could be attributed to boundary migration, in microduplex alloys the operation of this mechanism would require changes in composition across interphase interfaces. Then the difference in the composition of the constituent phases should be an important variable. Yet no major differences in the nature of superplastic deformation between alloys where the composition difference is a few percent, e.g. microduplex steels, or 100%, e.g. Bi-Cd eutectic, has been reported. Therefore boundary migration in the conventional sense is perhaps unimportant for superplastic flow.

(vi) There is some evidence for limited diffusional flow.

Although the low applied stresses and the high temperatures

of deformation should favour large scale diffusional flow, particularly in the vicinity of the grain boundaries, the mechanisms based on diffusion predict much lower strain rates compared with what are observed experimentally. Prediction of elongation of grains contrary to experimental results and the inability to explain the conversion of initially elongated grains to an equiaxed shape as a result of superplastic flow are additional difficulties. Moreover, conventional diffusional flow is not likely to be the rate controlling step in view of the non-Newtonian nature of superplastic deformation.

(vii) Texture existing prior to deformation appears to have insignificant influence on superplastic flow. Initially anisotropic structure becomes more random on deformation. The rate of texture removal varies with grain orientation and/or nature of the phase. Requirements of local compatibility between the grains also appear to affect this rate. When the starting microstructure was isotropic no new texture peaks were introduced by superplastic deformation.

(viii) Superplastic deformation perhaps does not involve appreciable dislocation activity.

The generality of superplastic flow independent of the chemistry and the crystal structure of the component phases, the retention of an equiaxed shape, high values of m and the absence of clear evidence for the motion of dislocations during superplastic flow support the above statement. The suggestion (Section 3.2.2(c)) that the greater stability of some texture components should be understood in terms of crystallographic slip, on the other hand, is to be regarded as equivocal at present.

(ix) Although the retention of an equiaxed grain shape can be explained in terms of multi-mechanisms, e.g. by combining diffusional flow with dislocation motion, the case for each of these models is rather weak. Further, multi-mechanisms should be invoked only if the results cannot be explained by a single mechanism (for further discussion see Chapter 4).

(x) The deformation process is such that the grains move uniformly and randomly and enable the retention of an isotropic structure. The central parts of the grains perhaps do not undergo any deformation.

In this viewpoint optimal superplastic flow is the result of random sliding of grains under the influence of the external stress and this process is accommodated by flow of the boundary regions in various directions on account of the internal stresses generated by sliding. This picture can also account for the occasional departure from randomness during superplastic flow. The observation of anisotropic flow when the starting structure is non-random can also be understood.

Surprisingly very few theoretical treatments are available that provide a clear physical picture for grain boundary sliding. If sliding is considered a unit process it requires subsequent accommodation, e.g. diffusional flow, dislocation motion. Alternatively, it may be looked upon as the result of the accumulation of atomic scale displacements in the grain boundary region. In this case continual accommodation of the boundary positions is inherent in the physical description of the unit of flow. These ideas have been fully developed in Chapter 4.

3.3 Fracture

The path of loading as well as the state of stress influence the fracture behaviour of materials. Therefore an idea about the cumulative damage resulting from different processes (which is not additive) is necessary for the construction of fracture maps.

The present knowledge regarding fracture in (the highly strain-rate sensitive) superplastic alloys is rather limited because most of the treatments of ductile fracture ignore the effect of strain-rate sensitivity. Thus an understanding of the fracture behaviour of this class of materials will be useful in (a) evolving broad-based criteria concerning the general problem of ductile fracture, (b) being alive to the problem of high temperature cavitation and intergranular embrittlement encountered during superplasticity, creep, fatigue, hot working, welding etc. with a view to developing industrially important superplastic alloys and high temperature materials and (c) examining if the strain-rate sensitivity of materials can be increased without decreasing their strength so that the fracture toughness may be enhanced.

In this Section both the experimental findings and the theoretical ideas concerning fracture in superplastic alloys will be covered as this topic is somewhat out of tune with the rest of the monograph.

Moreover, the explanations that have been offered for the origin of cavitation and fracture are only qualitative.

3.3.1 Macroscopic aspects

Fracture of superplastic alloys is generally classified into two types: (a) "ideal" fracture where the specimen necks down to a very fine point so that further deformation is impossible on account of the number of grains in the cross-section being extremely small and (b) "brittle" fracture in which specimens that exhibit practically no necking fail at 90° to the tensile axis. Very often in this case the fracture surfaces are macroscopically planar.

Elongation at fracture obtainable in case of the former can be predicted with reasonable accuracy (Section 2.6). The Zn-Al eutectoid and the Sn-Pb eutectic alloys are examples of materials that fail in this fashion. Using ternary eutectic alloys it has been demonstrated that for materials that fail ideally the number and the type of the interfaces present are immaterial as the various types of boundaries make similar contributions to superplastic flow and no problem exists in accommodating grain boundary sliding (183). Outside the optimal range, however, cavities are observed even in this category of materials, e.g. Zn-Al eutectoid alloy tested at low temperatures and/or high strain rates (365,366); Zn-0.4% Al alloy which exhibited "honeycomb" type of fracture at a strain rate for which m was about 0.05 (190); Sn-Pb-Cd and Sn-Pb-Zn ternary eutectic alloys deformed at strain rates greater than $\dot{\epsilon}_t^*$ in Fig. 3.2(b) (183).

The equation of Zhurkov (332) which predicts a temperature dependence for the rupture life of the Arrhenius form and an apparent activation energy for the rupture mechanism which linearly decreases with the applied stress breaks down at very low stresses and high temperatures of relevance to superplastic flow. Moreover, at present there is no way of taking into an analytical framework the experimental observation (266) that the rupture life of superplastic alloys increases with grain size.

In case of materials that fail in an ideal manner just before rupture extensive deformation of grains and the formation, growth and interlinkage of cavities are seen near the fracture surfaces. (This could be due to the fact that the strain rates in the vicinity of the fracture surfaces may pertain to region III.) The final failure is due to grains pulling apart at the boundaries (129,350).

Occasionally premature failure is caused by localised non-optimal microstructures, e.g. small areas of lamellar structure in hot worked Sn-Pb eutectic alloy (350), in an otherwise highly superplastic material. Very often, however, grain boundary irregularities and/or particles of undesirable shapes, sizes and properties that are present at the boundaries are responsible for cavitation and sudden breakage of specimens (56,257,258,260,261). In case of the latter situation cavitation at the grain boundaries is extensive, is prevalent in the entire gauge length of tensile specimens and is first observed at a very early stage of deformation. In case of materials susceptible to cavitation at large strains the true stress registers an increase with deformation (97). The final failure is brought about by the growth and interlinkage of cavities and is inter-crystalline. On account of the ultra-fine grain size the distance between cavities is relatively large and a significant proportion of the grain boundaries remains unaffected by cavitation. Thus not uncommonly the interlinkage of cavities around clumps of grains is observed. The fracture of superplastic copper alloys and some micro-duplex low alloy steels, for example, falls in this category.

The absence of cavitation in the grip regions of tested specimens implies that the phenomenon is deformation-induced. In general alloys that are prone to cavitation in creep, e.g. copper, iron, aluminium and nickel base alloys, also cavitate during superplastic flow. The rate and extent of cavitation, however, is system-dependent; rate of cavitation of copper alloys is greater than that of alloys of aluminium (80); the volume fraction of voids at fracture is approximately 10% in Cu-40% Zn alloy but is only about 2-3% in case of microduplex low alloy steels (367). When extensive cavitation is present both the elongation at fracture and the post-deformation properties (Chapter 6) are adversely affected, even though under conditions of superplastic deformation m is very high, e.g. m of the order of 0.7 is obtained in case of low alloy steels prone to cavitation (275). When cavitation is extensive the maximum in ductility is encountered at a strain rate for which the combination of the value of m and the extent of cavitation is optimal.

3.3.2 Microscopic aspects

Cavitation has been observed with the aid of optical (97,257), replica (271) and transmission electron microscopic (258,259,261,345) techniques.

In case of alloys that do not contain any particles at the boundaries, e.g. Cu-2.8Al-1.8Si-0.4Co alloy, binary and ternary alpha-beta brasses, the preferential sites for cavity nucleation were interphase interfaces, triple junctions involving (a minimum of) two phases, twin boundary-grain boundary intersections and less frequently intercrystalline boundaries of the harder phase (262,267,366,368). When denuded zones form as a result of superplastic deformation, e.g. in magnesium alloys, cavities could be found at these zones (238,345). In aluminium bronze 75% of the cavities were found at the interphase boundaries and triple points containing both the phases. The corresponding figures for the alpha-alpha and the beta-beta intercrystalline boundaries were 15 and 5 percent, respectively. The location of the remaining 5% of the cavities could not be clearly identified (262). In this category of alloys occasionally impurity particles of submicron dimensions are found associated with some of the cavities (367).

When hard and large particles were present at the boundaries either naturally, e.g. (about one volume percent of) Ti(C,N) or niobium/vanadium carbides in low alloy steels (103,275,276); cementite particles in plain carbon steels (271); iron-rich delta phase particles in alpha-beta copper alloys (56,260-262), or as a result of some artificial treatments given to materials that do not cavitate normally, e.g. introduction of about 5 vol.% of Ag_3Sn or 6 vol.% of Cu_6Sn_5 particles to Sn-Pb eutectic by the addition of small amounts of silver and copper respectively (367,369), cavities mostly formed at these particles. At large strains cavities were also found at the interphase interfaces, grain boundary obstacles, twin boundary-grain boundary intersections and/or triple points (102,276,368). However, it has been speculated that even in these latter cases particles could have been present above or below the planes of section (102,103,276). Moreover, there is some evidence that the extent of cavitation seen for a given experimental condition increases with the hardness of the grain boundary particles, e.g. introduction of the harder Cu_6Sn_5 particles gave rise to more severe cavitation compared with that of SnSb particles in case of the Sn-Pb eutectic alloy (367).

Cavities in general have a circular cross-section and often have a size equal to that of the grain size (129). Occasionally, however, cavities of elliptical cross-section with the major axis parallel to the tensile direction are seen. In this regard there is a speculation that the aspect ratio is maximum when the elongation to fracture is the largest (259,267). At triple points the cavities have a triangular shape. At interphase interfaces, on the other hand, they appear generally well rounded but have sharp apices (262). In some cases the cavities also lie in 'chains' along the stress axis (see below).

When hard particles are present at the grain boundaries the cavities are first encountered at very low strains of the order of 0.03 but become numerous only when a strain of about 0.4 is reached. Except in one case (258) where cavitation was considered to be independent of strain, in general both the number and the volume fraction of cavities increased with strain (97,262,276,370).

Depending on the alloy composition and the experimental condition the extent of cavitation for a given strain increases (258,259, 367), decreases (103,267,276) or goes through a maximum (97,370,371) with increasing strain rate. In some other cases the cavitation behaviour was essentially independent of strain rate (367,368,370). In some alloys, e.g. aluminium bronze containing 4% iron, no cavitation was seen below a strain rate of about $\dot{\epsilon}_t^*$ (Fig.3.2(b)) but at higher strain rates the severity of cavitation increased with strain rate (262). Even in materials that do not generally cavitate under optimal conditions cavities are often seen at strain rates greater than $\dot{\epsilon}_t^*$ (61,183).

In some cases the extent of cavitation increases with an increase in grain size and/or a decrease in temperature, e.g. alpha-beta brass, microduplex steels under some conditions (97,103). In others within the experimental range the susceptibility for cavitation was either practically independent of grain size and test temperature, e.g. low alloy steels deformed in a certain range (367,368) or increased with increasing temperature (103).

In one set of experiments (127) the (macroscopic) features of fracture were said to be unaltered when the flow stress was kept constant (while varying other parameters). In another (367) even when the strain rate sensitivity index and the elongation to fracture were unchanged the cavitation behaviour varied significantly with composition. In general the degree of cavitation decreased with an increase in the proportion of the softer phase.

When m was high void distribution was uniform but when its value was low cavities were mainly found in the necked regions. For constant m the severity of cavitation increased with an increase in the number of particles at the grain boundaries. Moreover, the nucleation and the interlinkage of cavities increased considerably just before fracture. On the other hand, factors that enhance superplastic tendencies, e.g. increase in m and/or test temperature; decrease in strain rate and/or grain size, decreased the tendency for cavitation (261,262, 270). Even when grain boundary embrittlement was deliberately induced by subjecting a microduplex stainless steel to nuclear irradiation the above conclusion remained valid (278). Occasionally small additions, e.g. titanium in case of a nickel-base Ni-Cr-Fe alloy (77); boron in case of low alloy steels, segregated to grain boundaries and decreased the extent of cavitation perhaps by lowering the rate of diffusion of vacancies to cavities (102,276).

Cavitation initially commences at boundaries normal to the stress axis although at large strains this process becomes random (190). A change in the strain rate of deformation, on the other hand, had the following effect on the cavitation behaviour of Cu-2.8Al-1.8Si-0.4Co alloy (371): (i) In region III of Fig.3.2(a) cavities were small and elongated and lay in the form of stringers parallel to the stress axis. Interlinkage of cavities was not common. Individual grains could be seen on the fracture surfaces only in isolated places (Fig.3.21(a)). Failure occurred by ductile fracture in which slip played an important part. (ii) In the strain rate range of transition from region II to III the cavities were larger, more randomly distributed and less markedly aligned. Their rugged appearance, on the other hand, perhaps indicated that the time available for the growth of cavities by diffusion was still insufficient. Although individual grains were visible over the entire fracture surface the final rupture still occurred by a ductile tearing process (Fig.3.21(b)). (iii) In the range of transition from region I to II (where elongation was maximum) the cavities were large, had a fairly uniform size and were randomly distributed. Individual grains were clearly visible and some evidence for the growth

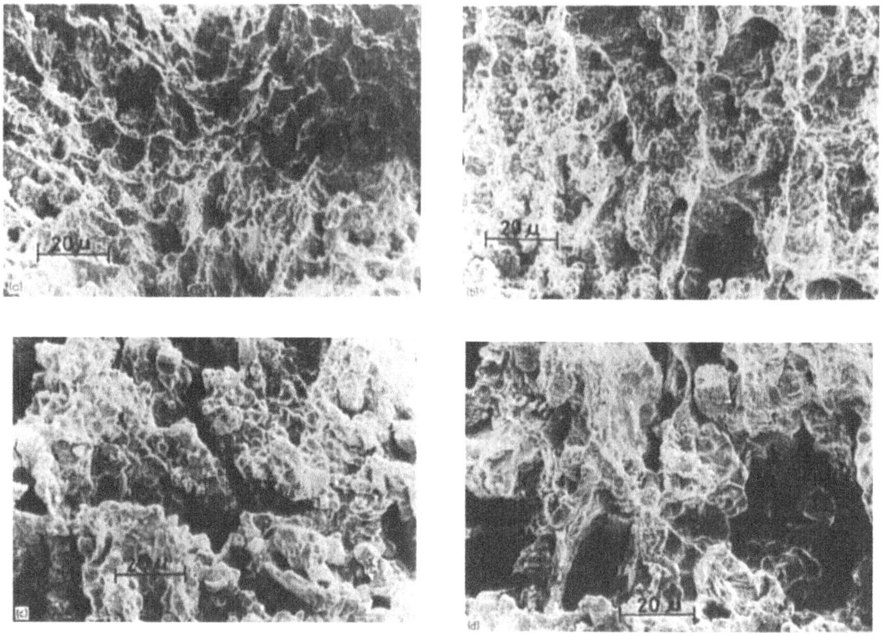

Figure 3.21 Scanning electron micrographs of the fracture surfaces of specimens of Coronze CDA638 (Cu-2.8Al-1.8Si-0.4Co) at different initial strain rates.
(a) $\dot{\epsilon}_t$=3.3x10^{-2} s^{-1}, (b) $\dot{\epsilon}_t$=8.3x10^{-3} s^{-1},
(c) $\dot{\epsilon}_t$=3.3x10^{-4} s^{-1} and (d) $\dot{\epsilon}_t$=1.7x10^{-5} s^{-1}
(Shei and Langdon (371))

and interlinkage of cavities was also found (Fig.3.21(c)). (iv) In region I the cavities were very large and well rounded. Some grain growth was also present. Final failure was due to interlinkage of cavities (Fig.3.21(d)). (v) In region I the cavities were large and numerous on account of limited boundary migration. (In alloys in which the migration rate is high, e.g. Cu-40% Zn, the number of cavities observed in this region is very small.) (vi) Void growth and interlinkage becomes increasingly important as the strain rate is decreased. (More specifically below a strain rate of about $\dot{\epsilon}_t^*$ the fracture is intercrystalline (100).) (vii) Even when the elongation to fracture at two different strain rates (say, in regions I and III) was similar the behaviour during cavitation and fracture varied widely.

Although the above conclusions have been drawn using a quasi-single phase material, earlier investigations (100,102,251,267,275,276) appear to indicate that they are applicable in case of multi-phase alloys also.

The above results were obtained using tensile tests. It has been reported that in case of the Cu-38.5%Zn-3%Fe alloy when the method of loading was changed from tension to compression no cavitation could be detected (97). The fracture behaviour in fatigue of superplastic Zn-Al eutectoid alloy prepared from commercial grade raw materials, on the other hand, was dependent on grain size (372). In specimens with a grain size of about 1.7 μm the β phase that had segregated during the casting stage and large (size $\sim$ 50 μm) Fe_3Al type inclusions (which result from the impurities present in the raw materials) act as crack nucleating sites. When a crack chain attains a critical length it becomes a fatigue crack. This crack then grows preferentially and final failure appears to be intercrystalline. When the grain size was coarser ($\sim$6 μm) the fatigue cracks grew only along the interphase boundaries. In this case the inclusions had no effect on the growth rate of cracks. Moreover, the fatigue life increased with frequency (which is a measure of the imposed strain rate). Evidently the high strain-rate sensitivity of the material decreased the amount of plastic deformation at the crack tip when the frequency was raised and this led to an enhancement of the fatigue life. The effect of frequency was retained down to ambient temperatures and was eliminated only when the ultra-fine microstructure was destroyed.

3.3.3 Origin of cavitation

(a) Phenomenological

Based on the hypothesis that fracture is the result of internal necking between adjacent microscopic cavities a theory of ductile fracture for a material with a strain-rate sensitive yield stress has been proposed for plane strain conditions (373). In the analysis necking occurs only at transverse rows of cavities. When the ratio (k_n/k_u), where k_n is the shear yield stress in the necked region deforming at a strain rate $\dot{\epsilon}_n$ and k_u is the shear yield stress corresponding to the uniform rate of deformation $\dot{\epsilon}_u$, exceeded a critical value superplasticity resulted. The degree of superplasticity, on the other hand, depended on the hydrostatic stress, the transverse stress and the volume fraction of the cavities. Moreover, consistent with experimental results, for materials with a finer dispersion of second phase particles a higher value of the (k_n/k_u) ratio and elongation to fracture were predicted.

Fleck et al. (267), however, consider the intergranular void sheet (IVS) mechanism to be important in case of the fracture of superplastic alloys and have suggested that the inequality

$$(1/\sigma_t)\ K\epsilon_t^N\dot{\epsilon}_t^m \times [(N/\epsilon_t) + (m/\dot{\epsilon}_t)(d\dot{\epsilon}_t/d\epsilon_t)] < (3/8)^{\frac{1}{2}}\left(\frac{1+z}{1-z}+1\right)^{\frac{1}{2}} \times F_{zb}2\left(\frac{2b^i}{1b^i}\right)^2 \quad .. \quad 3.10$$

where z is the void eccentricity, $F_{zb}2$ is the void growth factor, $2b^i$ is the initial horizontal width of the void, $1b^i$ is the initial horizontal spacing between the voids and the rest of the symbols have their usual meaning (chapter 2), defines the onset of fracture. Thus in this point of view the strain-rate sensitivity index as well as the void size, geometry and spacing control the fracture behaviour.

The onset of cavitation has been compared with the commencement of stage I instability during macroscopic necking (Section 2.5) while the beginning of cavity interlinkage is considered analogous to the start of stage II instability (97,370). According to Taplin and Smith (370) cavities that nucleate at grain or interphase boundaries grow mostly as a result of plastic strain accumulation. The main feature in this case is the high resistance to necking and failure offered by the inter-cavity ligaments. As the strain rate in the ligaments is considerably greater than the uniform strain rate or the rate of deformation at the external neck, the above somewhat unusual property can be understood by postulating that the ligaments possess strain-rate sensitivity. The mechanistic origin of this type of behaviour, however, is not clear.

(b) Microstructural

In general pure metals are less prone to cavitation than alloys. Therefore by controlling the purity of materials and by changing the diffusivities along the interfaces by appropriate additions (which in turn will control the rate of flow of vacancies to the cavities) the cavitation behaviour can be considerably altered.

It has been suggested (370) that when the superplastic alloy contains no particles at the grain boundaries the condition of stroh for nucleating a crack at triple junctions, viz. $\tau_c^2 > \frac{12\gamma G}{\pi L}$ where τ_c is the critical stress necessary for cavity nucleation, γ is the surface energy, G is the modulus of rigidity and L is the length of the boundary or the separation between the precipitates that block sliding, is not satisfied. If this is the case the segregation of trace impurities to the grain boundaries will control the extent of cavitation. In general, however, in this category of alloys (i.e. no particles are present at the boundaries) cavitation is considered to be the result of the inability of the accommodation process(es) to maintain boundary cohesion in presence of significant sliding. In particular, accommodation is difficult at interphase boundaries and triple points involving (a minimum of) two phases because (a) these are numerous when the volume fraction of the second phase is large; (b) on account of greater misorientation between the grains of dissimilar phases, i.e. of the interphase interfaces, both the rate and the amount of sliding could be greater; (c) critical energy required for the nucleation of a cavity at the interphase boundaries could be less compared with that needed in case of other types of boundaries; (d) when the phases have dissimilar properties accommodation could be easier on that side of the interface where the softer constituent is present and this can lead to an imbalance (of accommodation) and therefore cavitation; and (e) the stress concentrations present at the interphase boundaries and triple points involving two phases cannot be relieved by migration on account of the excessively large diffusion distances involved (119,261,276,368).

On the other hand, when hard particles are present at the boundaries cavitation usually results from the decohesion of the particle/matrix interface due to sliding. Moreover, thermodynamic considerations have revealed that the particles should have a minimum size in order to serve as effective sites for cavity nucleation (119). Occasionally, however, cavities nucleate at smaller particles and/or grain boundary irregularities. The ease of cavitation depends on the magnitude of the stress concentration which, in turn, is determined by the nature, number, size, shape and hardness of the particles as well as the strength and the propensity for precipitation of the particle-matrix interface.

Similar conclusions have been reached by Ridley and coworkers (102,103,276,367,368) by using the simplified Avrami equation (374), i.e. $V = Kt^n$ (applicable for small values of V) where V is the volume fraction of the voids, t is the time of deformation and K and n are experimental constants, for identifying the physical processes responsible for cavitation. The following assumptions are involved in this method: (a) the magnitude of n is independent of the composition of the alloy and the imposed strain and (b) a given range of value of n has the same physical significance in case of both phase transformation and cavitation. The analysis has revealed that when particles are present at the boundaries a substantial number of voids grow from pre-existing nuclei, probably found at the non-coherent particle-matrix interfaces. In quasi-single phase alloys, however, the results were consistent with a picture in which the cavities grow from nuclei that form at a high rate in the early stages of deformation.

The growth of cavities can result from either grain boundary sliding (when it is faster than the sintering rate of cavities by diffusion) or vacancy condensation mechanisms. In one case (367) it was not possible to identify the mechanism (out of the above two) relevant to superplastic flow. In general, however, based on the relatively large size of the cavities, their irregular shapes and the rather small values of the ($\sigma_t / \dot{\epsilon}_t$) ratio in the normal experimental range vacancy condensation processes are considered to be unimportant to superplastic deformation and therefore cavity growth is essentially the result of grain boundary sliding (97,119,370).

The dependence of cavitation on temperature, strain rate and grain size (Section 3.3.1), on the other hand, cannot be readily understood only in terms of lack of accommodation for grain boundary sliding (103). For example, an increase in temperature invariably increases the diffusivity and the boundary mobility. Yet in some cases the extent of cavitation increases with temperature (while in others it decreases). Likewise accommodation of sliding should be easier at lower strain rates but still in a number of experiments the void volume increased with decreasing strain rate. If boundary particles alone are important for cavity nucleation the cavitation behaviour should be independent of grain size. On the other hand, if lack of accommodation caused cavitation a strong dependence on grain size is expected. In practice, however, the grain-size dependence is rather weak.

The case where the cavitation behaviour is practically independent of strain rate and temperature has been explained by postulating that (a) virtually all nucleation of cavities occurs in the early stages of deformation, (b) its growth depends only on the rate and extent of sliding and (c) in the optimal range of strain rates and temperatures the contribution from sliding to total deformation is approximately constant. For this situation the volume fraction of the voids is only a function of the imposed strain (368). In this regard it should also be noted that acoording to a speculation of Dunlop et al. (262) the strain rate dependence of the cavitation process is a consequence of the change in the mechanism that accommodates sliding, e.g. from diffusion to dislocation motion with increasing strain rate.

Briefly, cavitation is a prelude to fracture. However, in some superplastic alloys it commences at a very early stage of deformation even when the experimental conditions are optimal. When this happens the decrease in the area of cross-section at fracture is rather small and the premature failure is due to the formation, growth and interlinkage of cavities. In case of systems in which there are no particles present at the boundaries the preferred sites for cavity nucleation are interphase interfaces, triple points containing a minimum of two phases, and twin boundary-grain boundary intersections. Less frequently cavitation is also present at intercrystalline boundaries of the harder phase and at denuded zones. When large (grain boundary) particles are present cavities predominantly form at the incoherent interfaces between the particles and the matrix. In general the extent of cavitation increases with strain. The effect of grain size, strain rate and temperature on the cavitation behaviour, however, is somewhat complicated and is system-dependent. The size, shape, number and distribution of the cavities that are encountered critically depend on the experimental conditions, the nature, number, size, shape and hardness of the particles (if present) and the alloy composition. Even when the alloys were prone to cavitation in tension no cavities could be seen in compression. The behaviour in fatigue, however, appears to be somewhat similar to that seen under conditions of tensile loading.

Phenomenologically the onset of cavitation is analogous to the start of stage I instability in case of external neck formation (Section 2.5) while the interlinkage of cavities is comparable to the occurrence of stage II instability. Regardless of whether particles are pre are present at the boundaries or not, it appears that the explanation for the commencement of cavitation and its subsequent growth is to be found in the inability of the accommodating process(es) to maintain boundary cohesion in presence of significant sliding. Detailed arguments that take into account the effect of experimental variables with a view to evolving a self-consistent framework, however, are lacking. Under the circumstances it seems reasonable to suggest that the present knowledge concerning fracture in structurally superplastic alloys is rather limited, at least in comparison with what is already known regarding their flow characteristics.

4. Structural Superplasticity – Theoretical

It was mentioned in Chapter 1 that early Russian workers attempted to explain the occurrence of structural superplasticity in terms of solution-precipitation reaction or the development of metastability. Underwood (4) has discussed these early theories in detail. In view of their historical importance these theories are briefly reviewed in Section 4.1.

As the number of alloys that are structurally superplastic increased, the early theories were found to be inadequate. Thus commencing from the mid 1960's a number of models have been proposed which explain structural superplasticity in the light of contemporary knowledge. These modern theories are based on either rheology (continuum treatments with or without the incorporation of microstructural details like grain size) or considerations that take into account the actual atomic structure of metallic materials. Depending on the aim of the analysis, the rheological approach can be considered in two parts (Sections 4.2.1 and 4.2.2), viz. (a) understanding the macroscopic response of superplastic materials to mechanical forces and (b) accounting for the measured values of the activation energy for flow.

Contemporary atomistic theories of structural superplasticity, on the other hand, can be divided into four sub-groups depending upon the (rate-controlling) step that determines the microstructural, topological and mechanical characteristics of flow. These are: (a) diffusion creep models, (b) modified dislocation creep theories, (c) grain boundary deformation models and (d) multi-mechanisms in which two or more independent processes contribute to the overall rate of deformation. In Table 4.1 the major analyses have been classified into the various sub-groups.

Many of the theories are speculative or make use of ideas originally developed for entirely different situations. A few models, however, are based on considerations fully relevant to superplastic flow and evidently these are of greater importance.

Table 4.1

Modern Theories of Structural Superplasticity

Type of Analysis	Sub-group	References	Section
Rheological (continuum)	Characteristics of macroscopic flow	Geckinli and Barrett (145) Gifkins (375) Hazzledine and Newbury (61, 377) Padmanabhan and Davies (72)	Section 4.2.1
	Activation energy for deformation	Langdon and coworkers (124, 293-295, 314, 378-380) Padmanabhan and coworkers (69, 319, 323, 334, 381)	Section 4.2.2
Atomistic	Diffusional flow mechanisms	Ashby and Verrall (62) Gifkins (170) Nabarro-Herring and Coble Creep (382-384)	Section 4.2.3(a)
	Dislocation creep theories	Ball and Hutchison (79) Chaudhari (33, 385) Hayden, Brophy and coworkers (77, 91, 279) Mukherjee (300)	Section 4.2.3(b)
	Grain Boundary deformation models	Beere (386, 387) Gifkins (388) Gittus (389, 390) Padmanabhan (63, 357, 381)	Section 4.2.3(c)
	Multi-mechanisms	Avery and Backofen (70) Baudelet and coworkers (82, 144, 304) Packer and Sherby (238, 239)	Section 4.2.3(d)

4.1 Early Theories

Bochvar (32) suggested that structural superplasticity is observed when solution-precipitation processes arising from the mutual solubility of the components lead to concentration fluctuations and formation of nuclei of a new phase. This idea was further developed by Osipov (391), Guy and Pavlick (see ref. (4)), Gurov et al. (392) and Novikov et al. (343).

The mechanism involves (system and composition dependent) mass transfer through the grain interior and/or along the interphase boundaries. Under the non-equilibrium conditions of deformation embryos of a second phase could form within the nearly saturated solid solution. This incipient phase transformation, in combination with the applied stress, leads to anomalous creep (by an unspecified mechanism) in the vicinity of the embryos. The change in the environment due to localised deformation, however, makes the embryos redissolve. A repetition of this process gives rise to cycles of precipitation and resolution and consequently to superplasticity. In this approach due to directional diffusion of atoms striated bands could form at the transverse boundaries and grow parallel to the stress axis. The absence of grain elongation has been attributed to grain rotation and switching (343).

Presnyakov and Chervyakova (38), Starikova and Presnyakov (40) and Presnyakov et al. (393), on the other hand, attribute structural superplasticity to the presence of metastable phases. The rate of deformation should be consistent with the rate of decomposition of the phases.

Partial décomposition (of the metastable phases) accompanies quenching or crystallization (when quenched from the liquid state). Segregation from the supersaturated solid solution also results during high temperature deformation due to the directional movement of atoms (on account of the applied stress). The rearrangement of the lattice in the latter stage weakens the interatomic bonds and facilitates deformation at very low stresses. The significant directional movement of atoms, on the other hand, leads to superplasticity.

In the quenched state the alloy is assumed to be made up of two kinds of metastable zones. On decomposition (during deformation) the zones transform into the equilibrium phases. Detailed analysis (40) led to the conclusions that (a) metastability was maximum when the volume fraction of the second component was 0.5 and (b) ductility increased with an increase in the proportion of the second phase.

When Underwood (4) wrote his review in 1962 he considered this approach to be useful because he erroneously attributed the positive influence (on superplastic behaviour) of an increase in the quenching temperature and the proportion of the second component as well as a decrease in the extent of homogenisation and the rate of deformation to the greater degree of metastability induced in the material. From the contemporary point of view, however, the first three treatments result in a finer microstructure while deformation at low strain rates ensures that there is no dislocation activity. Both these effects enhance the degree of superplasticity (Section 4.2.3(a)-(c)).

The above theories are of historical significance because they represent the first detailed attempts at providing a scientific explanation for the occurrence of structural superplasticity. They also rightly exmphasize the importance of interfaces and diffusion in understanding superplastic flow. However, as the number of superplastic alloys increased it was realised that hot worked and annealed specimens as well as alloys in which the decomposition of the metastable phases was complete well before the commencement of superplastic deformation, e.g. quenched and spinodally decomposed Zn-Al eutectoid alloy, could also be rendered highly superplastic. Moreover, the considerable mass of experimental results produced with the aid of carefully controlled investigations that employ powerful and diverse techniques (Sections 3.1 and 3.2) clearly demonstrate that neither precipitation-re-solution processes nor decomposition of metastable phases are of relevance to superplastic flow. It is also significant that the concept of break-down of metastable phases has been abandoned in the more recent papers published from the U.S.S.R. (see, for example, Kaibyshev et al. (131)). The balance of evidence reviewed in Sections 3.2.1 (b) and 3.2.2 (b), on the other hand, indicates that significant grain growth is absent during superplastic deformation. The small increase in grain size can be traced to limited diffusion, particularly because in situ electron microscopy (211,352) has shown that atoms of the right species get pulled out of grain neighbours and this facilitates the movement of boundaries.

4.2 Modern Theories

As mentioned earlier modern theories of structural superplasticity are based on either continuum mechanics (rheology) or atomistic mechanisms. In this Section the rheological theories are discussed first. This is followed by a review of the various analytical treatments concerning the activation energy for superplastic flow. The latter then serves as a useful parameter for confirming the validity of the atomistic models when they are considered towards the end of this Section.

4.2.1 Rheological theories

Constitutive equations based on rheological considerations are useful in those situations where knowledge about the operating atomistic mechanism is not necessary. Unlike in case of the atomistic models where the external stress is the independent variable, the above equations often assign this role to the strain rate. As optimal superplastic flow is encountered only within a narrow range of strain rate, obviously there are advantages in following the latter procedure. The basic rheological approaches are reviewed below. The applications of continuum mechanics in superplastic forming, on the other hand, will be taken up in Chapter 6.

Some of the rheological treatments are phenomenological while others incorporate microstructural details. This dichotomy of approach is consistent with developments in rheology where two sub-groups, viz. phenomenological and microstructural rheology, have already emerged (see, for example, Kroener (394)).

Hart (376) has suggested that superplastic flow could be represented in terms of a non-Newtonian viscous matrix (m less than unity) that contains a dispersion of plate-like flaws (equivalent to grain boundaries). Newtonian shearing (m = 1) occurs at the interfaces arising from the presence of the flaws. Combining the two types of responses in parallel gave a better correlation with data compared to when they were connected in series. At very low strain rates the flaws (grain boundaries) were assumed to offer no resistance while at high strain rates they were too strong to undergo deformation. The analysis has led to an equation

$$\sigma_t = \sigma_o \, (y - (1-y)\, Z) \; \dot{\epsilon}_t^{\,\mu} \qquad .. \quad 4.1$$

where σ_o and μ are phenomenological constants, y is the proportion of material deforming in a non-Newtonian viscous manner and Z is the ratio of the stresses carried respectively by the Newtonian and the non-Newtonian viscous elements. σ_o and μ depend on composition and temperature. y and another constant $\dot{\epsilon}_o$ (which correlates Z and $\dot{\epsilon}_t$), on the other hand, vary with μ and grain shape and size, respectively. The large number of adjustable constants present in the analysis and the assumption that at either end of the strain rate spectrum the matrix response rate controls (which should have been proved) considerably restricts the usefulness of this analysis.

A similar suggestion has been made by Hay and Pascoe (395) according to which regions I and II of Fig.3.2(a) result from the simultaneous operation of two interdependent processes with m = 1 and less than unity (= 1/3 for the Sn-Pb eutectic alloy), respectively. For ensuring geometrical compatibility the strain rates produced by both the processes have been assumed to be equal and of the same order as the external strain rate. The total applied stress, however, was the sum of the stresses needed to drive the two processes at the given strain rate. In this way m could be made to vary between (1/n) and 1 (where n is the stress sensitivity of the second process). The analysis further revealed that over limited ranges the log stress-log strain rate plots could appear linear and therefore the commonly adopted technique of extrapolating backwards the stress-strain rate curves to estimate a threshold stress (Section 3.1.1) for the commencement of superplastic flow is to be viewed with caution. The rather unlikely assumption that both the processes give rise to strain rates of equal magnitude, however, severely limits the scope of this analysis (129).

A greater number of mechanisms (four in all) and adjustable parameters (eight in number) have been used by Gifkins (375) to deduce the equation

$$\dot{\epsilon}_t = C \, L^{-(1+p)} \; \sigma_t^{\,n} \exp\,(-Q/RT) \qquad .. \quad 4.2$$

where C, p and n are constants and the other symbols have their usual meaning. In this approach equation 4.2 is the result of the combined action of (a) grain boundary sliding (n = 6, p = 0) accommodated by dislocation motion near the boundaries (n = 5, p = 0), (b) grain boundary diffusion ($n = 1$, $p = 2$) and (c) sliding accommodated by the motion of dislocations in the grain interior ($n = 2$, $p = 1$). Therefore, except at low stresses the characteristics of sliding were determined by those of the accommodation processes and appropriate choice of the operating mechanisms, on the other hand, ensured that both n and p assumed a range of values that included non-integral ones. In the process, however, even the activation energy becomes little more than an adjustable parameter.

The rheological model of Pompe and Feige (71) starts with equations concerning the thermally activated motion of dislocations. Both hardening and softening (due to dislocation climb) mechanisms are considered and an internal stress that is proportional to the mobile dislocation density has also been introduced. The predicted variation of the degree of superplasticity with the experimental variables was realistic. The authors, however, have conceded that the model was inconclusive with regard to the operating mechanisms. In its present form it (the model) is open to all the criticisms concerning the use of recovery creep theories and the concept of a back stress for understanding superplastic flow (see Section 4.2.3).

Rheological theories have also been developed by replacing the physical processes thought to be present during superplastic flow by their mechanical analogues. Zehr and Backofen (173) have combined in parallel a non-Newtonian "dash-pot" (grain boundary sliding) with another having a Newtonian response (Coble diffusion creep) and have then connected this combination in series to a third "dash-pot" displaying non-Newtonian behaviour (dislocation creep). In this approach there are four adjustable parameters and flexibility also exists in the way in which the total stress is distributed in the three "dash-pots". The diffusion data used allowed additional scope for adjustment. In spite of these possibilities the model predicted that the diffusional contribution decreased with decreasing strain rate (which was contrary to experimental results) (Section 3.2.1)(129).

Geckinli and Barrett (145) have adopted a similar procedure for the calculation of the stress reduction needed to achieve zero instantaneous strain rate. In this case grain boundary sliding and its accommodation at triple points and grain boundary protuberances are considered. When the stress is reduced, relaxation takes place by reverse grain boundary sliding. In the model a "dash-pot" representing diffusional accommodation at grain boundary obstacles is combined in parallel with a "spring" corresponding to elastic deformation at triple points. This combination is joined in series with another "dash-pot" representing diffusional accommodation at triple points. Assuming that the apparent viscosity arising from accommodation at grain boundary obstacles is one half of that due to the presence of the triple points, it has been estimated that the residual stress corresponding to zero instantaneous strain rate could be as high as two-thirds of the stress present at the beginning of the stress reduction test. Moreover, if the accommodation at the triple points and the grain boundary obstacles occurred by different mechanisms, e.g. volume and grain boundary diffusion, then a temperature dependence of the internal stress should also be expected. Evidently the internal stress was the result of the apparent viscosity of the grain boundaries which arose from the time-dependent release of the elastic strain energy and therefore should be distinguished from the threshold stress necessary for the commencement of superplastic flow.

In the model of Hazzledine and Newbury (61,377) grain boundary sliding is accommodated by the continuous deformation of grains (which is oscillatory with time) so that their equiaxed shape is always retained. Then the locus of a grain centre with initial coordinates x_o, y_o, z_o (during deformation) will be described by the equations

$$y = y_o\, e^{\epsilon_t}; \quad x = x_o\, e^{-(\epsilon_t/2)}; \quad z = z_o\, e^{-(\epsilon_t/2)} \qquad \text{.. } 4.3$$

if the stress acts along the y-axis. (e is the base of the natural logarithms.) When a set of grains lying on a plane z = constant are deformed the area of the plane will increase. As a result gaps appear between initially compact grains (Fig.4.1). But as the thickness (z) of the specimen decreases with deformation (equation 4.3) grains from adjacent layers (dark grains in Fig.4.1) fill these gaps. (At the surface, however, only one half of the total volume of the gaps is filled and this results in surface delamination.) A process of this type can continue indefinitely.

In order to ensure geometrical compatibility (396,397) it was assumed that the rate of grain strain was comparable to the sliding rate. It was suggested that the sliding rate was approximately equal

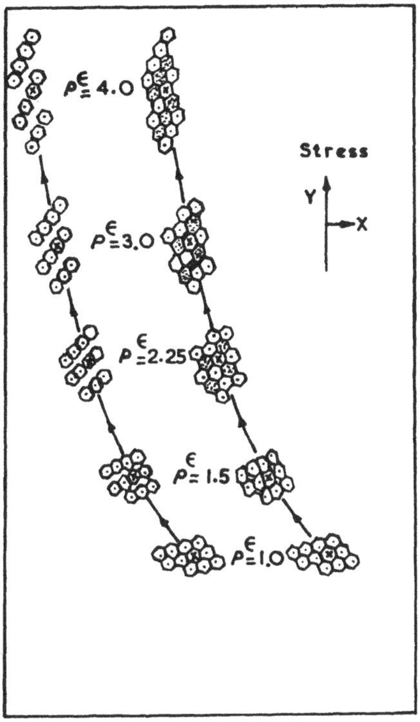

Figure 4.1 Grain boundary sliding and grain strain in a superplastic specimen pulled in tension along y-axis. Left: grain motion without accommodation; grains initially overlap but finally spread apart. Right: dark grains from adjacent Z planes fill the gap. (Hazzledine and Newbury (61))

to the ratio of the relative velocity between two adjacent grains to the grain size. The instantaneous rate of grain strain, on the other hand, was roughly equal to 0.625 times the sliding rate. Moreover, it was also assumed that small deviations from an equiaxed shape could increase the (grain boundary) surface energy and give rise to a back stress approximately equal to the surface energy to grain size ratio.

The grain strain is considered to be the result of the simultaneous and independent action of Coble creep, Friedel (dislocation) creep and Nabarro pipe (dislocation) creep. In addition intercrystalline boundary sliding is present. These four mechanisms pertaining to a single phase have been represented by four dash-pots. As superplastic alloys are usually microduplex, including the dash-pot which represents interphase boundary sliding a total of nine dash-pots are required in this approach. Moreover, contributions from interphase and intercrystalline boundary sliding are assumed to be equal and their magnitude was fixed from experimental results as no theory was found to be satisfactory. It was also assumed that both the phases possessed a similar mechanical response.

The analysis reveals that in region I of Fig.3.2(a) sliding (which is assumed to display a low strain-rate sensitivity) or the back stress (when this is large) determines the characteristics of flow. In region II diffusion mechanisms operate in parallel with grain boundary sliding and m of the order of 0.5 is obtained. Region III is due to dislocation motion but the fit in this region is rather bad perhaps due to the omission of dislocation glide from the analysis. Consistent with experimental results it has also been predicted that when the grain size is large Nabarro (dislocation) creep will dominate and superplasticity will be absent.

The large number of mechanisms invoked and the uncertainties associated with the constants involved in each of them restricts the use of this method. Moreover, the relation that the rate of grain strain is equal to 0.625 times the sliding rate may not be always valid. Finally for dislocation mechanisms the postulate that the total strain rate is the sum of the individual strain rates is perhaps not applicable (129).

At this stage it is relevant to observe that the significance of the operating mechanisms assumed in most of the rheological analyses reviewed above is rather doubtful. Moreover, the above procedures clearly illustrate how easy it is to fit mechanical data given sufficient number of adjustable parameters. Therefore it is essential to ensure that the microstructural and the topological features of a model are also consistent with experimental findings.

As a recent mechanistic model (63) (Section 4.2.3(c)) envisages superplasticity as a phenomenon arising from the (non-Newtonian) viscous flow of the boundaries of essentially non-deforming grains, Padmanabhan and Davies (72) have postulated that macroscopic superplastic deformation can be described in terms of a process in which a frictional force is generated at the grain boundaries because of the velocity discontinuity that exists there. Following Debye (398) this frictional force, which causes the grains to be dragged along with the moving boundaries, is assumed to be a linear function of the velocity discontinuity. (This assumption has been experimentally verified (72).)

The analysis has been developed for the case where the stress system is uni-axial (and acts along the x direction). Assuming that the velocity of all the boundaries is constant and on average equal to the external velocity of displacement, V, of the specimen, the x component of the frictional force acting on a grain of mass M whose centre has an instantaneous velocity V_x is represented as

$$F_x = f\,(V-V_x) \qquad .. \; 4.4$$

where f is a friction factor characteristic of the interface. (It has been experimentally demonstrated that f is approximately a constant for a given V.) The solution of the above equation is given by

$$x = Vt + (MV/f)\,e^{-(f/M)t} + x_o - (MV/f) \qquad .. \; 4.5$$

where t is the time of deformation and x_o is the initial x coordinate of the centre of mass of the grain under consideration.

It has been shown (using equation 4.5) that when the number of grains per unit volume is N, P_A is the volume fraction of phase A, f_A is its friction factor, M_A is the mass of an individual grain of phase A and the corresponding quantities for phase B are $P_B(= (1-P_A))$, f_B and M_B, respectively, then the stress is given by

$$\sigma_t = N^2V(P_Af_A\,e^{-(f_A/M_A)t} + (1-P_A)f_B\,e^{-(f_B/M_B)t}) \qquad .. \; 4.6$$

But

$$t = (\,(1/\dot{\epsilon}_t) - (l_o/V)\,) \qquad .. \; 4.7$$

where l_o is the initial gauge length of the specimen. As the mass of a grain is proportional to the cube of its linear size, then equation 4.6 (together with equation 4.7) expresses the stress in terms of strain rate and grain size (for isothermal deformation). A comparison of these equations with equations 3.1 and 3.9, on the other hand, will yield the value of the strain-rate sensitivity index.

If the distinction between the phases is ignored, it will

follow that $f_A = f_B = f$ and $M_A = M_B = M$. Then

$$\sigma_t = N^2 fV e^{-(f/M)((1/\dot{\epsilon}_t)-(l_o/V))} \quad .. \quad 4.8$$

$$= K_1 \dot{\epsilon}_t^{M_1} \quad .. \quad 4.9$$

where $K_1 = N^2 fVe^{(f/M)(l_o/V)}$ and $M_1 = -(f/M)(1/\dot{\epsilon}_t \ln \dot{\epsilon}_t)$. Equation 4.9 is analogous to the superplasticity equation 3.1. Experiments have revealed that the approximate equation 4.8 itself described macroscopic flow adequately.

It has further been shown that the viscosity increases with strain rate up to $\dot{\epsilon}_t = (f/M)$. Beyond this stage, however, the viscosity decreases with strain rate and superplastic deformation sets in. Thus this method correctly predicts the existence of a lower limiting strain rate for the onset of superplastic flow.

Unlike conventional continuum analyses this method takes into account the effect of grain size on flow stress. It also involves only a single empirical parameter, f.

However, the analysis assumes that the velocity of all the boundaries is equal to V (the external velocity of displacement). This is an approximation because the differences in the orientation of the boundaries and the nature of the interfaces can introduce variations in the boundary velocity. It has also been postulated that the total stress is a sum of the stress terms arising from the motion of individual grains. (This neglects possible interactions between grains.) Generalisations with regard to these aspects and the introduction in the analysis of three friction factors f_{AA}, f_{AB} and f_{BB} corresponding to the boundaries between grains of A-A, A-B and B-B, respectively, are yet to be done.

4.2.2 Activation energy for superplastic flow

Experimental evaluation of the activation energy for superplastic flow is a topic of controversy (Section 3.1.3). Briefly, the activation energy at constant stress, Q_σ, is obtained from a plot of log $\dot{\epsilon}_t$ vs 1/T or (1/m)log σ_t vs 1/T. The activation energy at constant strain rate ($Q_{\dot{\epsilon}}$), on the other hand, is estimated from the log σ_t vs 1/T relationship. Evidently Q_σ and $Q_{\dot{\epsilon}}$ are unequal and <u>apparent</u>. In the optimal range the temperature dependence of the activation energy is usually small but when this is significant a conventional Arrhenius plot will only result in a meaningless value.

Under conditions of optimal flow the activation energy is of the order of that for grain boundary diffusion but there are also a few cases where it has been reported to be close to the activation energy for volume diffusion. Frequently for the same system the magnitude of both Q_{σ} and $Q_{\dot{\epsilon}}$ varies with the experimental conditions. In general the apparent activation energy decreases with increasing stress, strain rate, temperature and/or strain-rate sensitivity index. There is also a case (134) where the dependence of Q_{σ} on the matrix grain size is direct and linear. However, when the structure is unstable during deformation the Arrhenius plots become non-linear.

Self diffusion, solute diffusion, stress-induced diffusion, volume diffusion and grain boundary diffusion could all contribute to superplastic flow although their relative importance will vary with the experimental conditions (61). An added complication can arise from the grain boundaries acting as imperfect sources or sinks for vacancies because in this case the activation energy will depend on the barrier that should be overcome before emission or absorption of vacancies can take place (399).

Analytical Procedures

In the method of Cahoon (86) an experimentally evaluated back stress (obtained by extrapolation) is subtracted from the applied stress and the logarithm of the effective stress is plotted against the inverse of the absolute temperature to obtain an activation energy. The assumed value of the back stress is considerably greater than that predicted by theory, e.g. Ashby-Verrall model (62), and therefore the calculated value of the activation energy has no physical/theoretical significance (400).

Langdon and coworkers have illustrated an alternative procedure using data pertaining to three alloy systems, viz. the Zn-22% Al eutectoid (124,293-295,378), the Sn-38% Pb eutectic (124,314) and the quasi-single phase alloy of composition Cu-2.8% Al-1.8%Si-0.4%Co (379,380). In this method the sigmoidal log stress-log strain rate curves are divided into three approximately linear portions corresponding to regions

I to III in Fig.3.2(a). An activation energy for each of the three regions is calculated using the "semi-empirical" equation

$$\dot{\epsilon}_t = (AGb/kT)\ (b/L)^p\ (\sigma_t/G)^n \exp(-Q/kT) \qquad .. \quad 4.11$$

where G is the modulus of rigidity, b is the Burgers vector, A, p and n are system-dependent, empirical constants and the other symbols have their usual meaning. For each range a constant but different value of n (= the slope of the experimental log strain rate-log stress plots in that region) is assumed. (The values of n for the single phase material were significantly greater than those used for the eutectic and the eutectoid alloys.) Within each region the activation energy was assumed to be independent of temperature and the stress level (of its evaluation).

The activation energy could now be estimated according to one of the following three methods:

(i) The temperature change method in which the appropriate quantities corresponding to two or three temperatures are substituted in equation 4.11 and the resulting simultaneous equations are solved to obtain the activation energy. (This method is commonly used in case of high temperature creep (401,402).)

(ii) Alternatively, log ($\dot{\epsilon}_t G^{n-1}T$) is plotted against the reciprocal of the absolute temperature, keeping the stress constant. This procedure is similar to what is usually employed in case of superplastic deformation to evaluate Q_σ, with the difference that the latter method does not take into account the variation in G with temperature. (But it has been suggested by Mohamed and Langdon (378) that for superplastic flow Q_σ (obtained without temperature compensation) is of the order of the real activation energy.)

(iii) A third method is to plot log ($\sigma_t^n/G^{n-1}T$) against (1/T), keeping the strain rate constant. (This method (without temperature compensation) was first introduced in the superplasticity literature by Ball and Hutchison (79).)

The mode of testing, e.g. tension, shear, the amount of deformation and the method of calculation did not affect the magnitude of the activation energy. In region I (Fig.3.2(a)) the activation energy was of the order of that for volume diffusion while in region II its magnitude was close to that for grain boundary diffusion. In region III the activation energy was similar to that for bulk diffusion in case of the eutectic and the eutectoid alloys but for the quasi-single phase material its magnitude lay between those for volume and grain boundary diffusion. No explanation by way of the operating mechanism in each of the three regions, however, has been provided.

The approach has the following limitations:

(i) Equation 4.11 is "largely empirical" and the rigidity modulus enters the equation to account for "dislocation interactions" (403). But dislocation models have limited use in understanding superplastic flow (Section 4.2.3(b)). Thus the choice of equation 4.11 to represent superplastic flow is questionable. Moreover, the "activation energy" determined from an equation that has <u>no</u> physical basis is a meaningless value even if fortuitously its magnitude is close to that of a physically significant parameter.

(ii) Microstructural evidence appears to suggest that similar mechanism(s) operate(s) below a strain rate of about $\dot{\epsilon}_t^*$ (Fig.3.2(b)) (1,61-63,68). Thus it may not be physically meaningful to assign two different constant values for n in regions I and IIa.

(iii) Within each region n has been assumed to be independent of temperature despite experimental evidence to the contrary (section 3.1.3). This leads to an overestimation of the activation energy in region I (400). In fact, according to Blum and Reppich (404) n could be considered to be (nearly) independent of temperature only over narrow ranges even in case of high temperature creep (where the dependence of n on temperature is seen to be considerably smaller).

(iv) Even when two interdependent mechanisms of comparable speeds operate a linear log stress-log strain rate relationship could be observed over certain ranges in strain rate (395). On the other hand, a recent theory (63) has predicted that n $(=\partial \ln \dot{\epsilon}_t / \partial \ln \sigma_t)$ varies linearly with the logarithm of the applied stress and/or the grain size as well as the reciprocal of the absolute temperature of deformation, even when a <u>single</u> mechanism is operating. Thus the division of the total range into three linear portions in order to suggest that a single mechanism operates within each region may not be realistic. The relevance of this argument becomes obvious when it is noted that the same data points (124) could be presented either as a smooth curve or as comprising three linear portions. (No sudden changes in the micro-structural and the topological features are seen at strain rates corresponding to regions I and IIa, respectively. Therefore, an abrupt change of slope between regions I and II is unjustified.)

(v) The linear correction applied to the low temperature data and the additional assumption made that the modulus of rigidity of the alloy is related to those of the constituent elements through the rule of mixtures could lead to sizeable errors.

(vi) The experimentally observed strain rates in region I are even greater than those predicted by the (grain-boundary diffusion based) Coble creep model (295). Yet, the activation energy for region I suggested by the analysis is that of bulk diffusion. This contradiction still awaits clarification.

(vii) When m = 1, from equation (3.9) $Q_\sigma = Q_{\dot{\epsilon}} = Q$, the real activation energy (319) (as this is the case of Newtonian viscous flow). However, for the analysis of Langdon and coworkers $Q_\sigma < Q$, when m = 1 (378). Further, it has also been suggested that the activation energy is independent of the stress level of its evaluation and that Q_σ and Q are nearly equal (378). Experimental findings, however, reveal a strong, (non-linear) inverse stress dependence of Q_σ as a result of which the latter displays a widely varying range of values (323).

Under the circumstances, it appears reasonable to start from equation 3.9 (keeping in mind that n(=(1/m)) is a variable) because this equation has both empirical (1,68,129,326) and theoretical (63) justification.

From definitions it could be shown (69) that for constant m

$$(Q_{\dot{\gamma}} / Q_{\tau}) = m. \qquad .. \quad 4.12$$

(Following standard practice in rheology stress and strain rate have been expressed in the shear mode.) Both Q_τ and $Q_{\dot{\gamma}}$, however, are apparent. In later papers (319,323,334) these quantities were related to the real activation energy (Q) for the flow process. The important conclusions were as follows:

(i) When $m = 1$, $Q_{\tau} = Q_{\dot{\gamma}} = Q$. when m is a constant less than unity (the case of high temperature creep), $Q_{\dot{\gamma}} < Q_{\tau} = Q$. Equation 4.12 holds good for both these cases. However, these ideal situations are not encountered during superplastic flow.

(ii) The superplasticity equation 3.9 is identical to its less commonly used alternative form (which has no theoretical basis but is the basis of evaluation of $Q_{\dot{\gamma}}$)

$$\tau \propto \dot{\gamma}^{m} \exp (Q'/kT) \quad .. \quad 4.13$$

only when m is unity. Therefore for the general case in which m is a variable it follows that

$$Q_{\tau} \simeq Q + kT^{2} \left((\ln a)/m^{2}\right) (\partial m/\partial T)_{\tau} \quad .. \quad 4.14$$

and

$$Q_{\dot{\gamma}} \simeq Q + kT^{2} (\ln c) (\partial m/\partial T)_{\dot{\gamma}} \quad .. \quad 4.15$$

where a and c are the (constant) stress and strain rate, respectively, at which Q_{τ} and $Q_{\dot{\gamma}}$ are evaluated (319). (Equations 4.14 and 4.15 become exact as $m \rightarrow 1$.) On both theoretical and empirical basis (ln a), $(\partial m/\partial T)_{\tau}$ and $(\partial m/\partial T)_{\dot{\gamma}}$ could be shown to be positive while (ln c) is negative. $(\ln a)/m^{2}$, on the other hand, is greater than $|\ln c|$ but $(\partial m/\partial T)_{\tau} > (\partial m/\partial T)_{\dot{\gamma}}$ (63,319). Therefore, the inequalities $Q_{\tau} > Q > Q_{\dot{\gamma}}$ and $(Q_{\tau} - Q) > (Q - Q_{\dot{\gamma}})$ hold good. Moreover, it could be concluded that the best estimates of the (real) activation energy are obtained from conventional Arrhenius plots when data pertaining to the range in which m is large (as close to unity as possible) are used. Further, $Q_{\dot{\gamma}} \rightarrow Q$ as $\dot{\gamma} \rightarrow 0$ (323).

(iii) It has been theoretically deduced (63,406) that in case of superplastic flow the activation energy is lowered by an amount $(b\tau^{n})$ in the direction of stress. As $\tau^{n} \propto \dot{\gamma}$ (1,68,129,326),

$$Q_{\dot{\gamma}} = Q_{\dot{\gamma}\,act} - k_{1}\dot{\gamma} \quad .. \quad 4.16$$

where $Q_{\dot{\gamma}\,act}$ is the value of $Q_{\dot{\gamma}}$ as $\dot{\gamma} \rightarrow o$ $(\simeq Q)$ and k_1 is a constant. Thus over limited ranges (within which n will be relatively constant) the plot of $Q_{\dot{\gamma}}$ against the strain rate level of its evaluation should be linear (323).

(iv) The isothermal viscosity of superplastic alloys varies as follows:

$$\eta = \eta_o \, |\dot{\gamma}|^{m-1} \qquad .. \quad 4.17$$

and

$$\eta = \eta_o^{(1/m)} \cdot |\tau|^{(\overline{m-1}/m)} \qquad .. \quad 4.18$$

where η_o is a reference state of viscosity. For the general case where m is a variable

$$(Q_{\tau} - (FQ_{\dot{\gamma}}/m)) = (kT^2/m)\left[((\ln a)/m)(\partial m/\partial T)_{\tau} \quad -F(\partial m/\partial T)_{\dot{\gamma}} \ln c\right] \qquad .. \quad 4.19$$

where $F = (1-\eta_o \dot{\gamma}^m (\partial m/\partial\tau)_T \ln\dot{\gamma})$. Evidently for constant m equation 4.19 reduces to equation 4.12.

(v) The dependence of Q_{τ} and $Q_{\dot{\gamma}}$ on temperature could be predicted for both constant and variable m (319). (This situation is encountered when the activation energy is evaluated over a very wide range of temperature.)

The above conclusions are well supported by experimental results. Equation 4.12 has been verified (see, for example, Maulik and Padmanabhan (323) for detailed comparison) in case of Sn-Pb eutectic, Pb-Sn-Cd and Pb-Sn-Zn ternary eutectics, Zn-Al eutectoid, Al-Cu eutectic and a microduplex Ni-Cr-Fe alloy (69,183,211,323). The results of Maulik and Padmanabhan (323), Langdon and coworkers (124,293-295,314, 378-380), Chaudhari (33) and Ball and Hutchison (79), on the other hand, reveal that both Q_{τ} and $Q_{\dot{\gamma}}$ decrease considerably on increasing the stress and the strain rate, respectively. This is a consequence of equation 4.16 (and the direct relation between stress and strain rate). The linear dependence of $Q_{\dot{\gamma}}$ on strain rate (over small ranges), predicted by equation 4.16 has also been verified (Fig.4.2(a)-(c)).

In the experiments of Hildebrand et al. (134) Q_{τ} increases linearly with the matrix grain size. This is because (when the other conditions are unaltered) the decrease in the magnitude of the activation

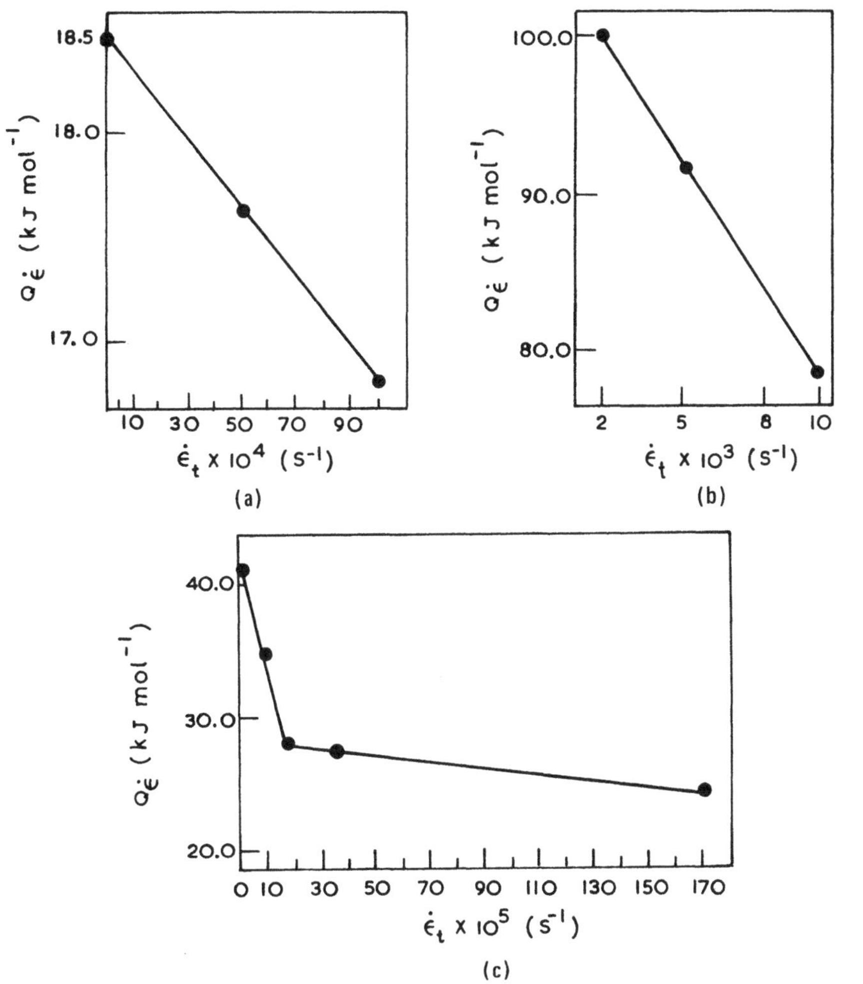

Figure 4.2 Plots showing linear dependence of $Q_{\dot{\epsilon}}$ $(= Q_{\dot{\gamma}})$ on $\dot{\epsilon}_t$ over narrow ranges in three systems (a) Sn-Pb, (b) Al-Cu and (c) Zn-Al (Maulik and Padmanabhan (323))

energy is inversely related to the grain size (Section 4.2.3(c)).

It was also clear that on an empirical basis the stress (or the strain rate) dependence of the apparent activation energy was significantly larger for superplastic flow compared with high temperature creep or hot working (69,183,323). Analyses other than the present one ignore this observation and could therefore lead to incorrect estimates of the activation energy.

Moreover, this analysis points out that when m is significantly lower than unity Q_τ is considerably larger than Q. For this case the ratio ($Q_\tau / Q_{\dot\gamma}$) is similar to that between the activation energies for bulk and grain boundary diffusion ($Q_{\dot\gamma} \simeq Q$). If the activation energy for superplastic flow is evaluated under these conditions (and Q is erroneously identified with Q_τ (77,100)) then its magnitude will be close to that for volume diffusion.

Consistent with theoretical predictions both Q_τ and $Q_{\dot\gamma}$ tend to Q as $m \to 1$. $Q_{\dot\gamma}$ also tends to Q as the strain rate approaches zero. In general, $Q_{\dot\gamma}$ is close to the activation energy for grain boundary diffusion although its magnitude always appears to be less than that of the latter (69,183,319,323). If the activation energy for grain boundary diffusion in the alloy is taken to be of the order of that for the fastest diffusing species, then Q_τ is considerably greater than the activation energy for grain boundary diffusion. If the activation energy of the alloy is considered to be the average of the respective values of the constituent elements, however, Q_τ is of the order of the activation energy for grain boundary diffusion (183).

Thus the balance of evidence suggests that for <u>optimal</u> superplastic flow the real activation energy is close to that for grain boundary diffusion.

In conclusion it is emphasized that the magnitude of the activation energy could provide confirmation about the rate-controlling mechanism that operates during superplastic flow. However, as two different processes can have similar values for their activation energy, speculations about the rate-controlling mechanism(s) solely based on the magnitude of the activation energy should be avoided.

4.2.3 Atomistic theories

It was seen earlier that modern atomistic theories of superplastic deformation can broadly be divided into

(a) Diffusional flow mechanisms,
(b) Dislocation creep theories,
(c) Grain boundary deformation models, and

(d) Multi-mechanisms that involve a combination of flow processes.

Many authors (see, for example, refs. 1,61-63,68) regard optimal superplastic flow over a range of strain rate, grain size and temperature to be the result of a unique mode of deformation in which the grain boundaries play a predominant role. Edington et al. (129), however, have taken a different stand on the following grounds:

(i) Several deformation processes, e.g. grain boundary sliding, dislocation motion, diffusion, have been reported in the literature as being present during superplastic flow. Moreover, in some systems the phenomenon is encountered at lower homologous temperatures compared with others. Therefore, there could be ranges of experimental conditions in which different mechanisms predominate and these (mechanisms) could produce by appropriate combinations the domain of superplastic deformation.

(ii) Experimental results are often inadequate for isolating individual mechanisms. Moreover, it is difficult to decide whether a mechanism is operating in its own right, as accommodation for a parallel deformation process or as the rate-controlling step.

(iii) In very few systems the relative importance of possible deformation modes has been assessed. When a number of mechanisms are operating it is difficult to decide as to how these should be combined to obtain an overall theory of deformation. This, therefore, means that the relationship between the mechanisms and the properties is somewhat uncertain and thus the relevance of mechanisms should not be judged merely by their ability to predict the properties accurately. Moreover, in many cases it is not possible to test the theories adequately because fundamental data, e.g. grain boundary or interphase diffusivity, are lacking.

Not withstanding these reservations it is evident that a more desirable point of view would be to attribute the phenomenon to a unique mechanism of flow provided that it can satisfactorily account for most of the experimental observations. Then it is possible to understand structural superplasticity on a common basis (63,68).

It is perhaps relevant to note at this stage that in the present Section a particular theory will be reviewed in that sub-section which is most appropriate to the development of the model, e.g. the Ashby-Verrall mechanism (62) dealing with diffusional accommodation of grain boundary sliding (in this model at no stage sliding is rate controlling) is reviewed in Section 4.2.3(a) on diffusional flow mechanisms.

(a) Diffusional flow mechanisms

Diffusion-controlled structural change and/or easy migration of vacancies are occasionally considered to be necessary for the occurrence of superplastic flow (42,138,221,306). The importance of diffusion is further reinforced by the observation that the addition of elements that enhance the diffusivity of alloys also gives rise to superior superplastic properties (101).

As superplasticity is a low stress, high (homologous) temperature phenomenon, both Nabarro-Herring (382,383) and Coble (384) creep could be important. For Nabarro-Herring creep the equation

$$\dot{\epsilon}_t = \frac{B_1}{L^2} \frac{\Omega \sigma_t}{kT} D_1 \qquad .. \quad 4.20$$

with Ω the atomic volume, D_1 the lattice diffusivity and B_1 a constant (approximately equal to 10 for materials with an equiaxed grain size), is valid. Thus for this model m = 1, the activation energy is of the order of that for volume diffusion and the strain rate is inversely related to the square of the grain size. Elongation of grains accompanies the flow process. For Coble creep, on the other hand, the approximate expression is given by

$$\dot{\epsilon}_t = \frac{B_2}{L^3} \frac{\Omega \sigma_t}{kT} w D_{gb} \qquad .. \quad 4.21$$

where w is the effective width of the grain boundary, D_{gb} is the grain boundary diffusivity and B_2 is a constant of the order of 150. In this case also grain elongation during deformation and a strain-rate sensitivity index of unity are predicted. However, the required activation energy is of the order of that for grain boundary diffusion and the strain rate depends inversely on the cube of the grain size.

Occasionally, both the above mechanisms are considered to be relevant for superplastic flow because the experimental scatter, the variations in m, the differences in the grain sizes of the constituent phases and grain growth and/or elongation prevent the identification of the operating mechanism based on the grain size dependence of the rate of flow (129,211,300,407). Nevertheless, there is a conclusion that under conditions of superplastic deformation Coble creep is faster than Nabarro-Herring flow by a factor of 10^5 (70,311). The balance of evidence suggests that both these mechanisms are inadequate for the following reasons (1,68,75,76,100,129,300,326,408):

(i) Evidence for the operation of these mechanisms in alloys is not conclusive. In any case, the Nabarro-Herring mechanism predicts strain rates that are too slow by several orders of magnitude. The Coble model, on the other hand, is claimed to predict the correct strain rates in case of some systems while for others the agreement is rather poor. Edington et al. (129) have, however, remarked that the agreement critically depends on the numerical values assumed for the grain boundary or the interphase diffusivity for which in general accurate data are lacking.

(ii) Contrary to experimental results in both the models the strain-rate sensitivity index is independent of strain rate and is equal to unity.

(iii) Both the models predict the elongation of grains and the retention of crystallographic texture existing prior to deformation, for which there is very little experimental support.

(iv) These mechanisms cannot account for the rotation of grains observed during superplastic flow.

(v) On account of grain elongation and the consequent increase in the diffusion path, these processes are self-exhausting and cannot give rise to superplastic strains.

(vi) At low stresses a linear dependence of stress on strain rate (m = 1) as well as high non-integral values of m (close to unity) could be predicted by using dislocation-based mechanisms also. Therefore, there is no need to invoke diffusion creep models merely to explain the high strain-rate sensitivity of superplastic flow.

The Harper-Dorn model with a strain-rate sensitivity index of unity is also considered to be unimportant for superplastic flow (409).

On account of geometrical constraints grain boundary sliding always accompanies diffusional flow (410,411). Moreover, sliding can occur without accommodation only at smooth, planar boundaries. Sliding at boundaries containing serrations, however, has to be accommodated by either slip or diffusion. For the latter case as well as for diffusional deformation, sliding and diffusional flow are inseparable. The simultaneous action of these two processes may be termed either as "diffusion creep" or as "grain boundary sliding with diffusional accommodation". In this case the (boundary) ledge height, h, its length, 1 and the repeat distance, λ , are the important variables (318,412, 413). The final expressions contain terms corresponding to both volume and grain boundary diffusion (see Edington et al. (129) for a detailed summary) and the two analyses available in this regard differ in their final results only by a factor of (λ/h). Volume diffusion is important only at high temperatures and/or when the ledge pairs are widely spaced. A change in the shape of the grains, on the other hand, can be avoided by confining sliding to a single system. An increase in the grain size or the aspect ratio and/or the presence of grain boundary migration in polycrystals considerably decrease the rate of sliding.

Some of these ideas have found application in the two-dimensional model of Gifkins (170) in which grain boundary diffusion is the rate controlling step. The superplastic system has been assumed to consist of regular, hexagonal grains. Rapid movement of vacancies and a uniform stress over each boundary are present. The average diffusion path has been taken as (L/4), where L is the grain diameter. If t is the thickness of atoms deposited at a boundary per second, the rate of diffusional accommodation is given by ($2t/\sin\theta$), where θ defines the orientation of the boundary with respect to the stress axis ($\theta = 60^{\circ}$ for hexagonal grains) and

$$t \simeq 56\, D_{gb}\, (\sigma_t\, w\Omega/L^2\, kT) \qquad .. \quad 4.22$$

The strain rates predicted by this method were too slow by a factor of about 10^5. Moreover, the following difficulties are also inherent in the analysis:

(i) The calculations are based on two-dimensional networks and no procedure for extending the analysis to the three dimensional situation has been provided.

(ii) The physical picture involves an intermediate, transitory change in the shape of the grains. In view of the remarkable similarity between the microstructures before and after superplastic deformation, models which do not require intermediate changes in the shape of the grains are preferable (342).

(iii) Contrary to experimental results, in this model the value of m is always unity (and independent of experimental variables like temperature, grain size and strain rate).

(iv) The speculation that grain boundary migration could also be an important accommodation step cannot apply in case of microduplex alloys in view of the excessively large diffusion distances that will be necessary.

The attempt of Gifkins (170) to combine the present results with those of an earlier study to predict that a grain size of about a micron is necessary for the occurrence of superplastic flow should also be viewed with caution because in the earlier model the rate of sliding linearly increases with grain size (while in the present case an inverse square relationship exists) and this type of dependence has not been reported in case of superplastic flow. In fact, in his later papers Gifkins himself appears to have abandoned this model (see Section 4.2.3(c)).

According to Backofen, Karim and coworkers (237,296-298,345) Coble creep and grain boundary sliding (which is a concomitant of diffusional flow) act in parallel and give rise to superplasticity. However, in this approach a threshold stress, σ_o, is necessary for the commencement of flow. Although no method of estimating the threshold stress has been provided, it has been speculated that this could arise from (a) the presence of rate-insensitive grain boundary shear resistance, (b) the action of boundaries as imperfect sources and sinks for vacancies and dislocations (see below) and/or (c) the need to activate grain boundary dislocations responsible for sliding. At low strain rates as σ_t is close to σ_o, m is low (cf. equation 3.9). But $m \rightarrow 1$ as $(\sigma_o/\sigma_t) \longrightarrow o$. Long before this stage, however, dislocation processes with intrinsically low values of m take over and m goes through a maximum with strain rate. But Lee (299) has remarked that in this argument σ_o has been used as an empirical, adjustable parameter because

(a) no method of predicting this quantity has been suggested and (b) experimental support for the presence of the threshold stress is not unequivocal (see also Sections 2.7.5 and 3.1.1).

More recently an effort has been made to trace the origin(s) of the threshold stress necessary for the commencement of "diffusion creep". As data concerning the presence of σ_o in superplastic alloys are lacking, arguments are necessarily based on results pertaining to creep (129).

When precipitates or particles are present at the boundaries commencement of diffusional flow may require a threshold stress. In case of polycrystalline ceramics cavities (introduced by sintering) could behave like precipitates and may necessitate the application of a threshold stress for the commencement of flow. However, it has not yet been established that the presence of a threshold stress is an intrinsic characteristic of diffusional flow in pure metals and ceramics which do not contain grain boundary precipitates, particles or cavities. Alternatively, if grain boundary sliding occurs along boundaries that contain some precipitates, a threshold stress may be necessary for its commencement. This idea is yet to be quantified.

A threshold stress may also be needed for the climb of dislocations in the grain boundaries containing particles or voids when this step is a part of "grain boundary sliding with diffusional accommodation". Here $\sigma_o = (G/b\lambda)$, where λ is the inter-particle spacing. Although Edington et al. (129) feel that this idea could be applicable in case of those superplastic alloys which contain some precipitates at the grain boundaries, evidently its relevance to a general theory of superplasticity is rather limited, as many (superplastic) alloys do not contain any precipitates at the boundaries. In any case, the concept has not yet been used in any theory of superplastic deformation.

By far the most popular explanation for the origin of the threshold stress has been in terms of the action of grain boundaries as imperfect sources and sinks for vacancies (interface reaction control). As at elevated temperatures there is an inverse relation between the strain rate and the grain size, the magnitude of the threshold stress could become significant when the grain size is very small. Physically this means that more time is taken over the creation and the annihilation of the vacancies than in their motion. The activation energy then will be equal to that needed for the emission or the absorption of vacancies (399,414). Alternatively, a threshold stress could arise from the need for the multiplication of grain boundary dislocations when this process is an essential part of the mechanism that gives rise to the creation and the annihilation of point defects (129). For example, the condensation or the emission of vacancies could occur by the glide-climb of dislocations in the grain boundary regions. Then, in order to maintain the density of grain boundary dislocations at an equilibrium value, dislocation multiplication will have to take place, say by the operation of a spiral climb source. For this case σ_o = (2G/bL). (Other dislocation configurations have been considered by Nichols (408) and Raj and Ashby (413).) The temperature dependence of σ_o in this case will be considerably less than that predicted by the former approach (i.e. boundaries act as imperfect sources and sinks for vacancies).

To-date the Ashby-Verrall model (62) (apart from that of Gittus-Section 4.2.3(c)) represents the only detailed analysis that provides a physical basis for the origin of the threshold stress in superplastic alloys. The development of this two-dimensional model (applicable in case of single phase materials containing equiaxed grains) differs from that of Nabarro-Herring and Coble creep mechanisms in topological details. In this case non-uniform flow leading to "grain switching" prevents significant grain elongation. Strain is produced by the relative translation of grains which also causes an increase in the number of grains along one of the two orthogonal directions. When the grains are not perfect hexagons the translation of grains will also produce a rotation. The switching of (grain) neighbours, on the other hand, will lead to large specimen elongation without altering the shape and the size of the grains significantly. Fig.4.3 schematically illustrates this grain-switching mechanism while Fig.4.4 features the diffusional accommodation that is necessary at the intermediate stage of the grain switching event (so that void formation at the boundaries is absent).

An idealised, two dimensional cluster of four grains constitutes the basic unit. As a result of flow, the shape of a group (of grains) as a whole changes but the shape of individual grains remains

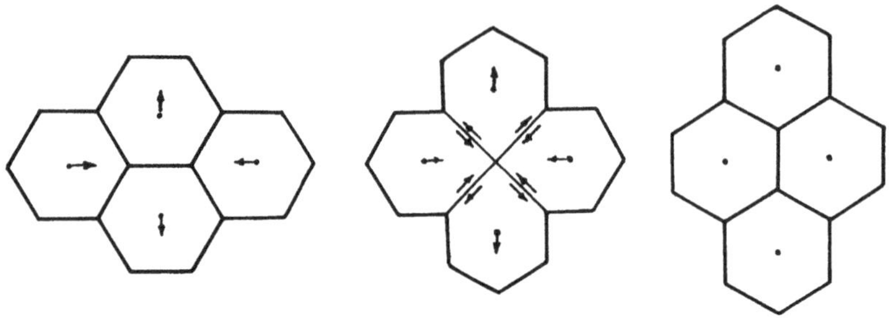

Figure 4.3 A schematic representation of the grain-switching mechanism of Ashby and Verrall. The directions of stress and relative movements between grains arrowed (Ashby and Verrall (62))

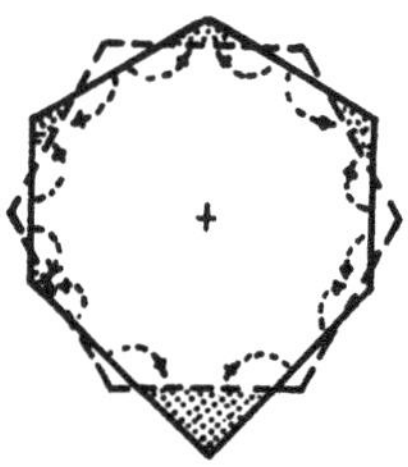

Figure 4.4 Diffusional accommodation at the intermediate stage of the switching event illustrated in Figure 4.3. The six diffusion paths have been indicated by arrows. The volume of material removed has been shaded. (Ashby and Verrall (62))

unaltered. The grains, however, suffer accommodation strain by grain boundary diffusion and this process prevents them from remaining together. Each unit step gives rise to a true strain of 0.55 and the change in the shape of a group of grains enables the external stress to do the work necessary for driving four irreversible processes:

(i) The diffusion process: A temporary change in the shape of the grains due to volume or grain boundary diffusion results because of the need for accommodation. If I is the total number of atoms or vacancies flowing per second from the sources to the sinks (within a group of four grains) and $\Delta\mu$ is the average chemical potential difference between a sink and a source, then the power dissipated by this process is $I \Delta\mu$

(ii) Interface Reaction: When the boundaries act as imperfect sinks or sources for point defects, an interface barrier has to be overcome before a vacancy can be removed from or added to a boundary. Although the magnitude of the chemical potential barrier could be different for the sources and the sinks, in the present analysis it is assumed to be equal to $\Delta\mu_i$ for both cases. Thus the power dissipated in driving the boundary as a source or a sink is equal to $2I\,\Delta\mu_i$.

(iii) Grain boundary sliding: Work has to be done against the grain boundary viscosity before sliding (shear displacements in the boundary plane which lead to relative translation between grains) can take place. The power dissipated by this process is $\tau A\dot{u}$, when τ is the resolved shear stress acting on a boundary plane of area A and $\dot{u}$ is the velocity of sliding. Power dissipated in this case is negligible although the sliding contribution here is greater than that obtained for Nabarro-Herring and Coble creep.

(iv) Fluctuations in boundary area: Energy is stored when a group of four grains changes from the initial to the intermediate configuration but gets released (as heat) once the original shape is restored. If Γ is the specific boundary free energy and $\Delta\dot{A}$ is the rate of change of boundary area, then the power dissipated by this process is $\Gamma\Delta\dot{A}$. Physically, this process gives rise to a small threshold stress below which flow is not possible.

When the boundaries act as perfect sources and sinks for vacancies the strain rate, $\dot{\epsilon}_{t\ \text{D-A flow}}$, for the "diffusion accommodated flow" process is given by

$$\dot{\epsilon}_{t\ \text{D-A flow}} = \frac{100\Omega}{kT\,L^2}\left(\sigma_t - \frac{0.72\Gamma}{L}\right)D_1\left(1 + \frac{3.3\,w}{L}\frac{D_{gb}}{D_1}\right) \quad .. \quad 4.23$$

This strain rate is greater than that predicted by the classical models, e.g. Nabarro-Herring or Coble creep, by a factor of about 7 because of the smaller amount of matter removed per unit strain <u>and</u> the shorter and the more numerous (six against four) diffusion paths. (The threshold stress ($0.72\,\Gamma/L$) results from the fluctuations in the grain boundary area.)

When the interface reaction rate controls, either an additional threshold stress will be superimposed on that due to the fluctuations in the grain boundary area or a strain-rate sensitivity index of 0.5 (413) will be predicted. Although the interface reaction is considered to be important for superplastic flow, this aspect of the problem has not been developed analytically.

Dislocation creep, which gives rise to a strain rate $\dot{\epsilon}_{t\ dis.creep}$ could compete with diffusion-accommodated flow. For the present, an equation

$$\dot{\epsilon}_{t\ dis.creep} = A\left(\frac{Gb}{kT}\right)\left(\frac{\sigma}{G}\right)^n \exp\left(-\frac{Q_c}{kT}\right) \qquad .. \ 4.24$$

where G is the modulus of rigidity, A is a constant, b is the Burgers vector and Q_c is the activation energy for dislocation creep, has been assumed. The total strain rate is obtained by adding the right hand side of equations 4.23 and 4.24.

In the low stress region more than 99% of the total strain rate is due to diffusion accommodated flow. In this range m is very low when the applied stress is close to the threshold stress. However, it increases with increasing stress and tends to a limiting value of unity. In this region grain elongation and formation of sub-grains and dislocation tangles are absent. Preferred orientation present in the starting material is gradully destroyed by sliding and grain rotation. At higher stresses, on the other hand, more than 99% of the total strain rate is produced by dislocation creep. In this case grain elongation, formation of cells and dislocation tangles and development of crystallographic texture are common. In the intermediate (stress) range where both the mechanisms are important, the behaviour of the superplastic material lies between the above two extremes. Mutual interference between the two types of flow processes, however, narrows the range over which both of them are important.

Consistent with the metallographic results reported in Section 3.2 (see also refs. 1,61,63,68) this model emphasizes that a single flow process is present in regions I and II(a) of Fig.3.2(b). It also successfully accounts for the sigmoidal variation of the log stress-log strain rate relationship. The agreement between this theory and the experimental results with regard to the microstructural and the topological features of superplastic deformation, on the other hand, are attributable to the nature of the assumptions made in the analysis. However, the Ashby-Verrall model cannot be applied to materials with a "bamboo structure" or in case of alloys with equiaxed grains when the strain is less than 0.55.

Garde et al. (111) have concluded that their experimental results are consistent with the predictions of the Ashby-Verrall mechanism. In general, however, the model has been criticized on the following grounds:

(i) Numerical difference between the predicted strain rates and the experimental values is rather large (293-295,314,379).

(ii) The model predicts that the strain rate is proportional to L^{-p}, where p varies from 0 to 3. Experimental results, however, reveal that p lies between 2 and 3 (close to 2 in the optimal range) and is often independent of stress and temperature (293-295,314).

(iii) In the model as the stress is increased n(=1/m) decreases from infinity to about unity. Experimentally, however, extremely high values of n are seldom obtained (293,294,314). Moreover, in many cases the predicted values of m are greater than those determined experimentally (415).

(iv) Contrary to experimental results the theory predicts that at a given temperature the stress-strain rate curves for different grain sizes will intersect (293,294,314,379).

(v) The prediction that the activation energy for flow increases with temperature has no experimental support (293,294,314).

(vi) In this mechanism the shape of the grains changes transitorily although the initial and the final configurations are similar. Models of this type are less attractive for explaining superplastic flow than those in which no change in the shape of the grains is required (342).

(vii) It is not clear from the model how the strain produced by the deformation of a unit of four grains is connected to those arising from other units. Nor has it been indicated as to how flow in a plane should be related to the deformation along the (third) orthogonal direction (61).

(viii) The experimentally observed motion of grains is not consistent with those predicted by the model, perhaps because of an erroneous assumption concerning the orientation of the grains with respect to the stress axis (308).

(ix) The model involves a number of physical parameters. With the aid of these and the choice of appropriate values for the threshold stress, considerable changes in the shape of the log stress-log strain rate curves and the magnitude of m can be brought about. The physical basis for these operations, however, would be questionable (73).

(x) The manner in which a non-linear stress-strain rate relationship has been developed in the model is rather sketchy. Moreover, the whole concept of interface reaction control is under debate (416). Other energy storage and dissipation processes, e.g. redistribution of the solute atoms under the applied stress, could also be important (129).

(xi) Langdon and coworkers (293-295, 314, 379) and Rai and Grant (60) have concluded that no clear experimental evidence for the existence of a threshold stress is available. Moreover, according to Smith et al. (415) (a) the predicted values of the threshold stress are too small to account for the experimental data satisfactorily and (b) if the existence of a threshold stress is assumed, then on an empirical basis its magnitude should decrease with increasing temperature (while in the model the magnitude of the threshold stress is independent of temperature).

Edington et al. (129) have further noted that the experimental evaluation of the threshold stress is done at very low strain rates and strains less than 0.1%. It is doubtful whether steady state flow conditions would have been reached at such small strains. Moreover, there are significant differences in the microscopic and the macroscopic features of deformation under these conditions and during optimal (region II of Fig.3.2(a)) superplastic flow. Therefore, the mechanism responsible for the existence of a threshold stress should be ascertained and shown to be operative in regions I and II (a) of Fig.3.2(b) before this concept can be considered to be relevant.

(xii) Quasi-steady state (grain boundary) diffusional flow has recently been examined in detail (416). It has been pointed out that continuity across the grains and the equality between the normal stresses acting on the opposite sides of a grain boundary are the conditions that should be satisfied in a realistic analysis of the problem. The following conclusions have been drawn: (a) Atomic movements envisaged by Ashby and Verrall for one of the four grains (in the unit of flow) are not permissible. and (b) For the assumed configuration of grains (62) there is no way in which the postulated atom fluxes can start. Therefore, the assumed diffusion paths are not correct. The intermediate configuration, on the other hand, cannot be reached even if grain boundary migration is present. (Moreover, although the reaction paths assumed by Lee (231) are consistent with the analysis, Ashby et al. (417) have correctly pointed out that in the former's model (231) the diffusion paths would be unrealistically large (because of the presence of grain boundary migration).)

In view of these criticisms, Ashby et al. (417) have proposed a modified, heterogeneous square grain model which employs shorter diffusion paths (compared with the model of Lee (231)) and achieves grain switching without boundary migration. Theorems for "power law" creeping solids have also been presented from which the bounding values of the strain rate (which should bracket the exact solution) for any model could be obtained. However, these theorems could not be applied either to the original Ashby-Verrall mechanism or to its modified version. Thus the relevance of both the modified, square grain model and the bounding theorems (417) for understanding superplastic flow is not clear at this stage.

In conclusion, therefore, it could be said that the case for each of the above diffusion-based mechanisms is equivocal. Nevertheless, it should be noted that the Ashby-Verrall model has been very widely discussed although in recent times it has been the subject of some critical statements.

(b) Dislocation creep theories

Recovery processes, particularly those confined to the grain boundary regions, could be important to superplastic flow (418,419). According to Kaibyshev and Kazachkov (420) (see Section 3.2.2(c) for a reference to the other papers published by these authors) superplastic deformation results from the combined action of non-basal slip, limited basal slip and diffusion. But Baero (421) regards the motion of dislocations and the sliding of grain boundaries as the two important mechanisms.

Naziri and coworkers (85,104,184,187,188) and Schmidt-Whitley (422,423) believe that at a given strain rate and temperature of deformation superplasticity will be present only when the sub-grain size is greater than the grain size of the alloy. Alternatively, the stress for which the cell size equals the grain size represents the upper limit of optimal superplastic deformation. (This approach, however, has been questioned by Young et al. (424).) In this case slip, dynamic recovery, grain rotation and sliding are considered to be important (189).

The absence of strain hardening and grain elongation, the development of an elliptical cross section in cylindrical tensile specimens and the changes in the intensities of the texture peaks (in superplastically deformed samples) have led to the speculation that crystallographic slip accommodated by grain boundary shear, migration and/or recrystallization leads to superplastic flow (192-194,289,290). Edington et al. (129), however, consider the case for recrystallization to be extremely weak (see also Sections 3.2.1 and 3.2.2).

In the opinion of Alden (166,322) superplastic flow results when the mode of deformation changes from slip to grain boundary sliding accommodated by slip and recovery. Briefly, region III(Fig.3.2(a)) arises from climb-controlled recovery creep and region II represents the resistance of the boundaries to sliding. Region I, on the other hand, is caused by the viscous glide of the dislocations created (by accommodating slip) at the triple points of the sliding boundaries. The absence of work hardening has been attributed to a balance between recovery and strain hardening.

It should be noted that the ideas reviewed till now in this sub-section have not been expressed in a quantitative manner.

Krausz (425-428) has associated the stress sensitivity (= $1/m$) of a superplastic alloy with a system of obstacles arising from dislocation mechanisms. In this approach Newtonian flow, the existence of an equilibrium rate between the breaking of atomic bonds and their mending and an interaction between plastic flow and the net rate of rupture of the bonds collectively give rise to superplastic deformation. Then,

$$\dot{\epsilon}_t = A_f \exp\left(\frac{n_f \sigma_t}{T}\right) - A_b \exp\left(\frac{-n_b \sigma_t}{T}\right) \quad .. \quad 4.25$$

where A and n are empirical material constants. The subscripts f and b refer to the forward and the backward reactions, respectively. The existence of four adjustable, empirical constants (which, in turn, depend on a number of physically ill-defined parameters), however, considerably restricts the usefulness of this analysis.

According to Chaudhari (33) the rate controlling step during superplastic flow is the motion of jogged screw dislocations in an internal stress field generated by the neighbouring dislocations. The driving force for the process is determined by the difference between the applied and the internal stresses. The latter stress has been assumed to be proportional to the square root of the density of mobile dislocations, ρ. If the shear rate of deformation is equal to ρbv, where v is the average velocity of the jogged screw dislocations, then the applied shear stress τ is related to the other variables by the equation

$$\tau = q\ \dot{\gamma}^{1/g} + S\ \dot{\gamma}^{(\overline{g-1}/2g)} \quad .. \quad 4.26$$

where $q = M^{-1/g} \exp(Q/gkT)$, M is a constant of proportionality, $S = \frac{\delta K_1 G(kT)^{1/2}}{(8\pi D \lambda_q)^{1/2}}$, δ is a factor that connects ρ with those dislocations giving rise to the internal stress, K_1 is a constant, g is another constant of the order of 4 or 5, λ is the average spacing between the jogs, D is a diffusivity term and Q is an activation energy. The constants of equation 4.26 for which numerical values could not be assigned were evaluated from the experimental results. Subsequent attempt at predicting the experimental results of Backofen et al. (42) using these values, however, was unsuccessful.

In the later dislocation cascade mechanism (385) Chaudhari represents grain boundary sliding (which is considered analogous to a viscous shear crack) by a double pileup of dislocations. At low strain rates diffusion creep at triple points controls the rate of flow and in the analysis the speed of this step has been equated to that of dislocation climb along the grain boundaries. If d is the length of the edge of a hexagonal grain in a regular two dimensional array and s is the sliding contribution, then an amount of material equal to (sd) should be removed by climb. For this case the strain rate is given by

$$\dot{\epsilon}_t \propto \frac{D_{gb}}{d^3} \frac{\tau_e \Omega}{kT} \quad .. \quad 4.27$$

where τ_e is the stress at the tip of the pileup. When the latter exceeds a critical value τ_c , however, the triple points release a cascade of dislocations. Then an increase in the rate of deformation (through a decrease in the climb distance) results. The increase in strain rate because of the cascade process is of the order of $(N+1)^2$, where N is the number of additional pileups produced. When each triple point gives rise to two pileups and the stress resulting from the cascade process is equal to the square of the applied stress, then the increase in strain rate will be about 10. It has also been claimed that the experimentally observed sigmoidal variation of the log stress-log strain rate plots is consistent with the predictions of the model.

Both the above models of Chaudhari predict the extensive formation of dislocation pileups within the grains. The absence of experimental evidence for the existence of dislocation networks, pileups and cells <u>and</u> for work hardening, when taken together with the very low flow stresses of superplastic alloys, provides a strong argument against this approach. In fact, experiments have clearly revealed that an increase in dislocation activity in the interior of the grains suppresses superplastic tendencies. The retention of an equiaxed grain shape after extreme (superplastic) elongations is also inconsistent with the predictions of these analyses. Moreover, the earlier calculation (33) uses three adjustable parameters and is therefore regarded by Edington et al. (129) to be a curve fitting exercise. Chaudhari himself has conceded that the assumption that the internal stress is proportional to the square root of the mobile dislocation density is unrealistic. The assumed correlation between the applied stress and the dislocation velocity is also tentative. Finally, for many systems the predicted separation between the jogs was greater than the grain size (429)!

According to Hayden et al. (77) a climb-controlled dislocation (pileup) mechanism based on the Mott (430,431) - Friedel (432) approach controls the rate of superplastic flow. As in the earlier treatments, the climb distance is taken to be of the order of the grain size and independent of the applied stress. The activation energy for climb, however, is assumed to be equal to that for lattice diffusion (Q_1) in view of the very small number of the dislocation pileups. Then,

$$\dot{\epsilon}_t \propto \frac{\sigma_t^2}{L} \exp(-Q_1/kT) \quad .. \; 4.28$$

At higher stresses, however, the climb distance varies inversely as the square of the applied stress and in this range m=0.25.

In a later paper (91) equation 4.28 has been modified as

$$\dot{\epsilon}_t = \left(\frac{A_1 b^3}{T}\right) D_1 \frac{\sigma_t^2}{L^2} \quad .. \; 4.29$$

(where A_1 is a constant that incorporates the geometric and the modulus terms) by making the following assumptions: (i) The number of active dislocation sources is about one per grain. (ii) The climb distance is proportional to the reciprocal of the applied stress. (iii) The grain boundaries can emit or absorb vacancies and contain a random array of vacancy-saturated jogged dislocations. and (iv) A parallel diffusion process with a diffusion distance of (L/4) is present.

Hayden, Floreen and Goodell (279), on the other hand, have taken the diffusion distance to be equal to (10b). The existence of a back stress (σ_o) has also been assumed. Below a temperature T_c, dislocation "pipe" diffusion or grain boundary diffusion (no distinction between the two processes is made as they have similar values for their activation energy) is the rate controlling step. Above T_c volume diffusion controls the rate of flow. The strain rate for the two cases then is given respectively by

$$\dot{\epsilon}_t = \frac{6\pi(1-\nu)\, D_1 b^2 \sigma_t(\sigma_t - \sigma_o)}{10\, GL\, kT}; \; T > T_c \quad .. \; 4.30$$

and

$$\dot{\epsilon}_t = \frac{2\pi(1-\nu)\, D_p b^3 \sigma_t(\sigma_t - \sigma_o)}{L^2\, G\, kT}; \; T < T_c \quad .. \; 4.31$$

where ν is Poisson's ratio and D_p is the "pipe" diffusivity along a dislocation core. The lower strain-rate sensitivity of materials with a fibrous microstructure, on the other hand, is predicted by multiplying the right hand sides of the above equations by a term proportional to ($\sigma_t L$). In addition it has been proposed that at the lowest stresses vacancy creep is the rate controlling step.

On account of their similarity it is appropriate to discuss together the above three theories of Hayden and coworkers. The following deficiencies have been noted (129,235,429):

(i) The predicted strain rates are too slow by a factor of about 10^6 to 10^8, particularly when volume diffusion is the rate controlling step.

(ii) The predicted inverse relationship between the strain rate and the grain size (equation 4.28) has no experimental support.

(iii) m is not always equal to 0.5 for superplastic deformation.

(iv) Clear evidence for dislocation activity during superplastic flow is absent. In any case, single dislocation-based models cannot explain the retention of an equiaxed grain shape, the randomization of texture and the presence of considerable grain boundary sliding. Moreover, when the number of active dislocation sources is one per grain the von Mises criterion is violated (433) and it is not clear from the models how slip on a single system should be combined with diffusion and sliding to ensure coherence of deformation across the grains.

(v) The models imply that either grain elongation or recrystallization should take place. Experimental support for neither is present.

(vi) Clear experimental evidence for the presence of a threshold stress (equations 4.30 and 4.31) is absent. In any case, no procedure for theoretically estimating this quantity has been outlined.

(vii) The assumption that the activation energy for flow in region II (of Fig.3.2(a)) is of the order of Q_l is not borne out by experimental results (314).

(viii) In the third model (279) the constant of proportionality connecting the strain rate due to sliding with that due to dislocation creep has been estimated arbitrarily for a grain size of 150 μm (73).

(ix) Although it is possible to alter with the aid of a number of ill-defined physical parameters and back stress the shape of the log stress-log strain rate plots significantly, the physical basis for the operation should be regarded as rather weak (73).

In the model of Ball and Hutchison (79) sliding accommodated by dislocation motion leads to superplastic flow. Protrusions on the boundaries of certain grains obstruct easy sliding between groups of grains and therefore the former (type of grains) yield under the stress concentration. The dislocations traverse the grains and pile up against the opposite boundaries until the back stress (due to the pileup) prevents further sliding. Dislocation climb into and along the grain boundaries makes further sliding possible. Fig.4.5 is a schematic

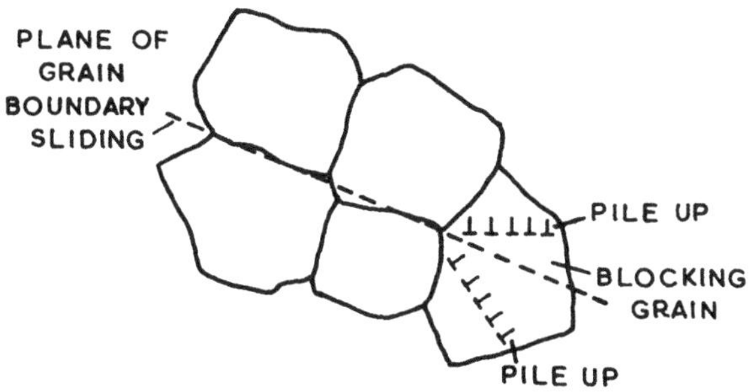

Figure 4.5 A schematic representation of the model of Ball and Hutchison (79). Sliding between a group of grains leads to rate-controlling slip in the blocking grain. (Edington et al. (129))

illustration of the above model.

The rate controlling step, therefore, is the climb of dislocations in the boundary region which requires an activation energy Q_{gb}. Grain rotation accompanying sliding continually changes the most favourable plane for slip and prevents grain elongation. The mechanism, however, will cease at stresses high enough for the formation of cells or dislocation tangles (because in this case the dislocations cannot traverse the grains).

If R is the ratio of the number of easily sliding to obstructing grains, (L/x) is the climb distance (x is a constant) and the climb velocity, v, is a linear function of the grain boundary diffusivity and the vacancy concentration, then it is possible to correlate the climb velocity to the applied stress through an intermediate evaluation of the local stress at the head of the dislocation pileup. As the time of sliding, t, is equal to (L/xv) and the strain rate is approximately equal to (a/Lt) (a is the unit of sliding),

$$\dot{\epsilon}_t = \frac{2a\ x^2\ b^2\ R^2\ \sigma_t^2}{L^2\ G\ kT} D_{ogb} \exp(-Q_{gb}/kT) \qquad .. \quad 4.32$$

Based on their experimental results Ball and Hutchison suggest that R=4 and x=2.

Mukherjee (300) has slightly modified this analysis by postulating that the grains do not slide together as groups and that the climb distance is of the order of L. For this case

$$\left(\frac{\dot{\epsilon}_t\ kT}{D_{gb}\ Gb}\right) = K'\ (\sigma_t/G)^2 (b/L)^2;\ K' = \frac{2a}{b} \simeq 2 \qquad .. \quad 4.33$$

Below a value of $(\sigma_t/G) \lesssim 10^{-4}$, however, Coble creep is considered to be the rate controlling step. At high stresses, on the other hand, conventional dislocation creep with n(=1/m) of the order of 4.3 determines the characteristics of flow. Vaidya et al. (209) have also drawn similar conclusions.

Schmidt-Whitley (243) and Arieli and Rosen (286) have concluded that their results are consistent with the models of Ball and Hutchison and Mukherjee. (No preference, however, has been indicated in view of their similarity.) Matsuki and coworkers (253,344), on the other hand, have preferred the model of Ball and Hutchison.

In general, however, these two mechanisms have the following limitations (203,209,269,293,294,315,380,429):

(i) The model of Mukherjee predicts strain rates that are more than an order of magnitude too slow when ocmpared with the experimental results.

(ii) A value of m=0.5 is not unique in case of superplastic deformation. Often m also depends on the experimental variables like grain size, strain rate and temperature.

(iii) Both the models do not account for the observed sigmoidal variation of the log stress-log strain rate relationship.

(iv) The model of Ball and Hutchison involves the movement of groups of four grains. Experiments, however, reveal that the grains move individually, particularly when the structure is isotropic and the strains are large - i.e. under conditions of optimal superplastic flow.

(v) The models do not apply to two phase materials because in general the yield behaviour of the component phases will be different.

(vi) The absence of work hardening, recrystallization and sharp texture peaks in superplastically deformed specimens and the very low stresses required for deformation are inconsistent with dislocation mechanisms. Moreover, both the theories predict the formation of dislocation pileups in a number of grains for which there is no experimental support. Even when the specimens were quenched under load no pileups were seen. The existence of absorbed dislocations at the grain boundaries or the interphase boundaries of a superplastically deformed specimen has not so far been reported. Thus climb into or along the grain boundaries is unlikely to be the rate-controlling step. Moreover, in the model of Mukherjee the ledges move as a result of sliding and therefore the dislocations do not start on the same plane. Thus it is not clear how a pileup can be produced in this case (129). (see also sections 3.2.1 and 3.2.2).

(vii) According to Stowell (429) the relationship $\dot{\epsilon}_t \propto \sigma_t^2$ can be predicted only if the glide systems of each grain are assumed to be non-interacting. Once the dislocation density is large enough to form pileups n(=1/m) should be of the order of 3 or 5, as in conventional dislocation theories.

Alden (322,434) has noted that the dislocation theories are derived from one of the two alternative approaches, viz., (i) a process which describes the rate of loss of dislocations by climb and (ii) the recovery creep model which relates the steady state creep rate with the rates of recovery and strain hardening.

In the dislocation pileup models the climb distance is proportional to the grain size and rate of climb depends on grain boundary diffusion rather than on volume diffusion. When the density of the dislocation sources is constant, the strain rate is directly related to the square of the grain size. Physically this means that the dislocation sources exist in the interior of the grains and that the creep rate decreases with an increase in the number of obstacles (grain boundaries) to crystallographic slip. Evidently this type of analysis is not applicable in case of superplastic flow.

On the other hand, when the sources of dislocations are the edges of sliding grains and the number of sources per grain is constant, it is possible to predict a strain-rate sensitivity index of 0.5 and an inverse grain size dependence of strain rate. (The earlier derivation of Hayden and Brophy (91) which led to an equation in which the strain rate was inversely related to the square of the grain size is considered to be intrinsically contradictory.) If, in addition, the jogs on the climbing dislocations are assumed to be saturated with point defects, then the models of Ball and Hutchison and Mukherjee are derived. This latter assumption, however, is questionable in view of the very short diffusion distances (434).

When bulk diffusion is sufficiently rapid, however, the dislocations climb towards each other and a strain-rate sensitivity index of the order of (1/3) is predicted. The flow process requires an activation energy of the order of Q_1 and the strain rate is independent of grain size. This situation is encountered during conventional creep.

Notwithstanding these remarks Edington et al. (129) have concluded that dislocation motion is a viable mode of deformation in view of their results concerning texture in superplastically deformed alloys. However, it has already been pointed out in Section 3.2.2(c) that their interpretation of the texture results is equivocal.

Moreover, the other predictions of the dislocation theories like the occurrence of primary creep, unloading transients, strain hardening and considerable dislocation activity cannot be reconciled with the available experimental evidence concerning superplastic flow. Therefore the usefulness of the dislocation climb theories seems to be rather limited (434).

(c) Grain boundary deformation models

The mass of experimental results reviewed in Sections 3.2.1 and 3.2.2 appear to indicate that the mechanisms of superplastic flow based on grain boundary sliding are by far the most viable (1). Often sliding is postulated as a unit step which requires subsequent accommodation. In most of the analyses, however, the accommodating step is rate controlling and the characteristics of flow are determined by the mode of deformation assumed for this stage. Very few treatments are available in which the sliding process itself is rate controlling.

Bulk or grain boundary diffusion; grain rotation; grain rotation and diffusion; grain rotation, switching of (grain) neighbours and diffusion; grain boundary diffusion and recrystallisation; slip and boundary migration; grain boundary migration or recrystallisation; slip and dynamic recovery; and plastic flow due to dislocations nucleated at ledges etc. which move either along the grain boundaries or through the lattice by a combination of glide and climb are some of the accommodating processes that have been considered. Superplastic flow has also been suggested to arise when there is a change in the accommodating mechanism from dislocation motion to diffusional flow as the rate of deformation is lowered (62,78,90,94,111,158,162,165,166,174,175,213, 226,231,241,249,256-258,260,261,284,287,313,315,336,350,364,435-437). In none of these cases sliding is the rate controlling step at any stage and these models have already been discussed in Sections 3.2, 4.2.3(a) and 4.2.3(b).

Beere (386,387) has pointed out that the significant grain rotation observed in region II (Fig.3.2(a)) of superplastic deformation can be explained only when the creep rate is controlled by either the shear process (grain boundary sliding) with rapid relaxation of the normal forces or the simultaneous slow relaxation of both the normal and the shear forces. In both cases the important feature from the point of grain motion is the effect of the shear forces. The friction which develops between the grains leads to their rotation. Thus creep rate-controlled by grain boundary sliding could be responsible for optimal superplastic flow. The above idea has been applied to the situation in which a polycrystalline aggregate made up of cubic grains is deformed in tension. Accommodation of the sliding displacements is confined to the grain boundary region and is achieved by rapid diffusion, i.e. the normal boundary stresses relax very rapidly. Then the aggregate will be supported only by the boundary shear stresses and the much slower relaxation of the latter will determine the creep rate.

According to this picture, deformation is confined to a "mantle" on the outside of the grains in whose interior no shear takes place. The physical details of the shear mechanism, however, have not been specified. Instead, a phenomenological equation $\dot{\gamma} = K_1 \tau^n$, where

$\dot{\gamma}$ is the shear rate, τ is the shear stress and K_1 is a constant, has been assumed (for the description of the shear process). Subsequently, the shear stresses appearing on the faces of the cubic grains and the associated motion of the grain centres have been calculated. When the cubes were assumed to deform by the path of least energy expenditure, the shear stresses acting on the cube faces were automatically balanced. When n was taken as 2 and the condition of coherence of deformation was imposed, the analysis led to the conclusion that the grains roll over their neighbours and give rise to grain rotation as well as switching of (grain) neighbours. Retention of a nearly equiaxed structure could also be ensured.

The geometric features predicted by the model (386,387) are consistent with experimental observations (Sections 3.2.1 and 3.2.2). It has also been emphasized correctly that the significant grain rotation seen in region II of Fig.3.2(a) can only result when sliding is rate controlling. Experimental support for the presence of the physical processes invoked in the analysis, viz. grain boundary sliding and limited amounts of rapid diffusion, is also unequivocal. Finally, the rates of grain rotation calculated on the basis of this model are in good agreement with those measured experimentally in region II. It should also be noted that an atomistic model (63) (presented shortly before the above analysis of Beere) has likewise suggested that sliding is the rate controlling step during optimal (regions I and IIa of Fig.3.2(b)) superplastic flow (see later in this sub-section).

However, the treatment is phenomenological and the physical process responsible for sliding has not been identified. The stress sensitivity (=1/m) has been assumed to be equal to two (or in general a constant) and this unique situation is not encountered in the entire range of optimal superplastic deformation.

In a subsequent paper (438) a <u>non-unique</u> solution for the problem of grain switching has been presented with reference to a two dimensional hexagonal network containing a defect (a five and seven sided grain pair). The shape of the grains was controlled by minimising the surface tension. The glide of dislocations, centred on the five and seven-sided grain pair, was examined with the aid of a computer model. The effect of sliding and diffusion (both of which are important during superplastic flow) on the grain shape, however, was neglected.

Therefore, the physical processes incorporated into the computer model do not appear to be important for superplastic deformation. No procedure for extending the analysis to the three dimensional situation has also been outlined. Thus, grain switching by the mechanism considered in the paper (438) is not perhaps relevant to superplastic flow.

Gifkins (388) has suggested that a typical grain could be divided into a central "core" in which single crystal behaviour is present and a peripheral "mantle" where the presence of the grain boundaries modifies the pattern of flow. The widths of both the core and the mantle have been assumed to change with the experimental conditions. The behaviour of the mantle is considered to be extremely important in case of fine-grained materials and its width decreases as the experimental conditions become increasingly favourable for grain boundary sliding. Fig.4.6 is a schematic illustration of the model. In this

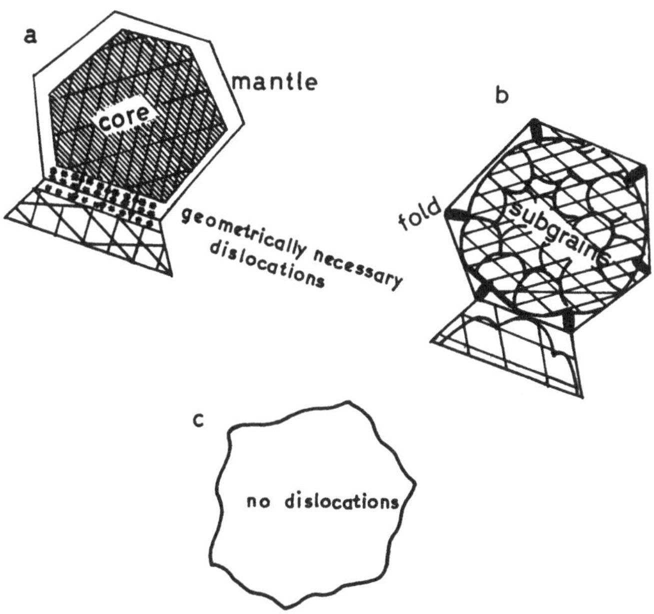

Figure 4.6 "Core and mantle" model of Gifkins for the deformation of a polycrystalline specimen. Effect of stress(or strain rate) on the nature and proportion of the two zones is indicated. (Gifkins (388))

case so long as the grain size is stable during flow the number of phases present is immaterial.

In region I (of Fig.3.2(a)) the rate of sliding is controlled by the barriers inherent in the structure of the grain boundaries. Ignoring the interaction between the neighbouring dislocations and assuming the (boundary) microfacets to be equally spaced, a semi-empirical equation

$$\dot{\epsilon}_t = (Z/L^2)\ \sigma_t^{1/m} \exp\ (-Q_I/kT) \quad .. \quad 4.34$$

where $Z = (2\beta/p)(db\nu)$ with β a constant, p a weighting factor less than unity, d the average distance between two nearest features on the grain boundary and ν the atomic frequency. The activation energy for deformation in region I has been taken as equal to $0.2Q_1$ and for the experimental verification of the model m has been assumed to be unity.

In the lower range of stress in region II diffusion creep via the grain boundaries controls the rate of sliding and an equation

$$\dot{\epsilon}_t = 82\ b^3 D_{gb}\ \sigma_t w/L^3\ kT \quad .. \quad 4.35$$

is obtained. (This is similar to the equation derived by Raj and Ashby (413) except for the value of the numerical constant.) As shown earlier (413) it is also possible to include in the above equation a term arising out of lattice diffusion.

In the upper range of stress in region II dislocation motion in the mantle accommodates sliding. Sliding is caused by the motion of grain boundary dislocations which pile up at the triple edges. Under the resulting stress concentration they dissociate into crystal lattice dislocations. Their subsequent glide and climb along the adjacent boundaries could cause grain rotation. Eventually they are assumed to either get annihilated or recombine to form new grain boundary dislocations. Under the circumstances, the interaction between the above dislocations as well as the (sparsely spread) dislocations present in the "core" could be neglected. Therefore, work-hardening will be absent. As this mechanism also is confined to a narrow mantle it is considered equivalent to that prevalent in the lower range of stress in region II. (Therefore, in the entire range of region II grain switching by a mechanism similar to that proposed by Ashby and Verrall (62) could be present.) The strain rate in the (upper) range is predicted by equation 4.33 when the numerical constant is changed from 2 to 64.

In region III sliding is accommodated by the motion of dislocations generated at the triple point folds. Crystallographic slip inside the core is also present. In this region the strain rate is given by

$$\dot{\epsilon}_t = \dot{\epsilon}_w (1 + 2F_y a/L^2) \qquad .. \quad 4.36$$

where $\dot{\epsilon}_w = 5x10^{20} D_1 b^3 \sigma_t^{4.5}/G^{3.5}$ kT, $F \simeq 2.3x10^{-2}$ cm when L is greater than the sub-grain size, a and $F \simeq 1$ cm when $L < a$.

The model has been subjected to the following criticisms:

(i) The experimental verification of the model is done essentially by curve fitting. Consequently the numerical agreement should not be regarded as a good test of the model (129). In any case, it has been noted by Gifkins (388) that the (numerical) agreement in each region is not very satisfactory except when different combinations of the flow processes are invoked.

(ii) Evidently, a model involving the multiple combination of mechanisms can predict a sigmoidal log stress-log strain rate curve, non-integral powers for the grain size dependence of strain rate and a spectrum of values for the activation energy ranging from Q_1 to Q_{gb}. This, however, need not mean that the mechanisms are in fact occurring. Independent direct evidence is lacking (129). (See also Sections 4.2.1, 4.2.3(a) and 4.2.3(b).)

(iii) It is not clearly stated why sliding should be rate controlling in region I (129).

(iv) The assumption that Q_I is equal to $0.2Q_1$ has no experimental support (379,380).

(v) The widths of the core and the mantle are used as adjustable parameters (that depend on the experimental conditions).

(vi) It has been suggested that the mechanisms operating in the lower and the upper ranges of stress in region II are equivalent (388). Yet, m=1 for the lower range while it has a value of 0.5 in the upper range. The microstructural features that will help distinguish the lower range from the upper one have not been spelt out. In fact, according to Edington et al. (129) direct experimental evidence for the presence of the mechanism of the upper range (of region II) is absent.

(vii) The model does not account for the retention of specific components of texture after deformation in region II (129).

Gittus (389,390) has presented a model for superplastic flow in which the Raschinger grain switching event (which is also used by Ashby and Verrall (62)) constitutes the unit step. The analysis, on the other hand, resembles that of Ball and Hutchison. The mechanism is applicable to duplex structures in which the phases are equally deformable, e.g. some eutectic and eutectoid alloys.

By postulating that the formation of superdislocations during the deformation of interphase boundaries (possibly by the expansion of intrinsic dislocation loops under the applied stress) could be a general phenomenon, sliding at these boundaries has been "provisionally" attributed to the glide of superdislocations (in the boundary plane). The dislocations are effectively trapped in the interphase boundaries and their motion produces (on these boundaries) dynamic (dislocation) pileups of length L (where L is the grain size), whose lead dislocation is located near a triple point. A gradual build-up of the (interphase boundary) superdislocations towards the equilibrium density, on the other hand, will make the boundaries "bulbous". At the triple point the interphase boundary (which can be described in terms of dislocations) meets two disordered boundaries into which the dislocations of the pileup will climb (the rate-controlling step). Then the rate of flow is governed by the equation

$$\dot{\epsilon}_t = 4.8(L/\lambda)^2(D_{gb}\ Gb/kT)(\sigma_t - \sigma_o/G)^2(b/L)^2 \qquad .. \quad 4.37$$

where the climb distance λ is proportional to the grain size, L. The threshold stress σ_o arises from the interaction between the interphase boundary superdislocations and the boundary ledges. When the ledge-dislocation network is replaced by a two-dimensional dislocation network and the width of the broad face of a ledge is d, then $\sigma_o \leqslant (2E/bd)$.

Loss of superplasticity, on the other hand, has been attributed to conventional dislocation creep which commences when the cell size is equal to the grain size. If (a) λ is permitted to range from

0.19L to 0.45L, (b) the length of the ledges is assumed to be about 100 times their height and (c) the expression for the threshold stress originally derived for the Ni_3Al-Ni_3Nb interphase boundary is considered to be valid for the Sn-Pb eutectic alloy, then satisfactory agreement between the predictions of the theory and the experimental results could be obtained.

This approach has the following limitations:

(i) The derivation is based on a two dimensional grain switching event which does not permit the emergence of new grains at the surface (439).

(ii) The model fails to predict either the increase in the activation energy or the decrease in ductility observed in region I of Fig. 3.2(a) (439).

(iii) The analysis applies specifically to two phase materials and as such is not able to account for the very similar observations made in single phase and quasi-single phase superplastic alloys (439).

(iv) Gittus himself has noted that the concept of a threshold stress is approximate. Even at stresses less than σ_o finite displacements at slow but finite strain rates will result (390). Therefore, it is not clear whether the existence of a threshold stress could be incorporated into a theoretical analysis, although its usefulness in a curve-fitting exercise is obvious.

With special reference to superplastic deformation, attention is drawn to Section 4.2.3(a) wherein it has been pointed out that (a) clear experimental evidence for the existence of a threshold stress is absent (60,293-295,314,379) and that (b) the experimental evaluation of the threshold stress is carried out at very low strain rates and strain less than 0.1% (and the microscopic and the rheological features of deformation under these conditions are quite different from those observed during optimal, steady state flow). Therefore the mechanism responsible for the presence of a threshold stress should be shown to be operative in regions I and II(a) of Fig.3.2(b) before this concept is used to explain superplastic deformation (129).

(v) The lowest value of the stress sensitivity predicted by

the model is 2. Therefore, values of m greater than 0.5 (reported in many systems) cannot be accounted for. Likewise, when the applied stress tends towards the threshold stress, the stress sensitivity (n) will tend to infinity. Such extremely high values of n are also seldom reported.

(vi) The postulate that sliding at the interphase boundaries is due to the glide of superdislocations (resulting from the expansion of intrinsic dislocation loops under the applied stress) is questionable (see below).

At this stage, it is appropriate to discuss briefly the present views regarding the nature of grain boundary sliding and the structure of high-angle grain boundaries.

Grain boundary sliding occurs at high homologous temperatures ($\gtrsim 0.4\ T_M$) in a variety of materials, e.g. metals and alloys, ceramics, rocks, ice, metallic glasses, polymers. In the literature two types of sliding have been identified (61): (a) "Lifshitz" grain boundary sliding which arises from the need to maintain continuity across the grains during diffusional flow and (b) "ordinary" grain boundary sliding which is encountered at strain rates considerably faster than those predicted by the common diffusional flow models. When the former type of sliding occurs, intragranular deformation is seldom reported but the latter type is often accompanied by crystallographic slip in a zone adjacent to the grain boundaries (405). At elevated temperatures sliding is easier than crystallographic slip (440).

The activation energy for sliding has been reported to be of the order of that for grain boundary diffusion in some systems while for the others its magnitude is equal to that for volume diffusion (405). A more recent survey (129), however, has concluded that the former value is more realistic.

The experimental techniques available for determining the contribution from sliding are controversial (Section 3.2.2(a)). Nevertheless, it could be safely said that the importance of sliding increases with decreasing stress and/or grain size (405).

The grain boundary region is of poor atomic fit and contains a high concentration of vacancies (197). High-angle grain boundaries contribute more to sliding compared with boundaries of small misorientations and in the limit it is conceivable that the atoms have complete freedom to move and prevent the formation of voids (315,342). During optimal superplastic flow in microduplex alloys, on the other hand, the sliding contributions that result from the three types of boundaries are comparable, although small differences do exist (1,196).

In general, the mechanisms of sliding are based on either diffusion or dislocation motion and there is a school of thought that both Lifshitz and ordinary grain boundary sliding is the result of the same physical process(es) (61).

Roy and Ranganathan (441) have noted that Mott's island model for the structure of high-angle grain boundaries (442), which pictured the boundary as consisting of islands of good atomic fit separated by seas where the fit is bad, has been followed by the coincidence-site lattice model (443) and the more refined "O-lattice" theory of Bollmann (444). In both the latter models rotations leading to partial lattice coincidence give rise to low energy boundaries and the deviations from these special orientation relationships are accounted for by superimposing dislocation arrays (referred to as intrinsic grain boundary dislocations (445)), which form an integral part of the boundary. In addition, crystal lattice dislocations and extrinsic grain boundary dislocations (which are caused by crystal lattice dislocations but whose Burgers vectors are not characteristic of the lattice) could also be present at the boundary.

Lin and McLean (446) consider the crystal lattice dislocations at the boundaries to be immobile. Gleiter et al. (447), on the other hand, have postulated almost all the behaviour of crystal lattice dislocations for extrinsic grain boundary dislocations, e.g. Frank-Read source action. According to Ashby (448) both intrinsic and extrinsic grain boundary dislocations can cause sliding, but grain boundary migration is necessary in order to sustain it. (This approach can only be of limited use in understanding superplastic flow- Section 4.2.3(a).) Except for some special orientations, e.g. pure twist, sliding will require both glide and climb of dislocations (440,449,450). Often the rate of sliding is controlled by dislocation climb (449) or the negotiation of obstacles such as crystal lattice dislocation/extrinsic grain boundary dislocation nodes (451). Slide hardening could result from the lattice dislocations providing pinning points at the grain boundary (449). The point defect source/sink action of the grain boundaries and diffusion creep have also been attributed to the motion of grain boundary dislocations (452).

On account of the above speculations sliding has been attributed variously to the motion of intrinsic grain boundary dislocations (448,449), crystal lattice dislocations (440,450,453-460) or extrinsic grain boundary dislocations (447,448,451,461-463). The picture is very confusing and it is yet to be established that sliding is caused by the motion of any of the three types of dislocations mentioned above (464). Moreover, except when diffusion is regarded as the rate controlling step the models have not been developed quantitatively (61). It is also significant that (to-date) in none of the superplastically deformed specimens absorbed dislocations have been observed at either the grain boundaries or the interphase interfaces (129).

Even when crystal lattice dislocations were initially present at the boundaries, annealing at a temperature well below the range in which significant sliding occurs made them disappear (465-468). Moreover, Misra et al. (469) have noted that the coincidence-site lattice model becomes less applicable as the symmetry of the crystal decreases, e.g. in hexagonal close packed structures very few coincidence-site lattice relationships have been established. Therefore, the interpretation of grain boundary sliding in ceramics in terms of a dislocation model will be difficult (as the latter have a predominantly non-cubic structure) (470).

In fact, recently the belief that the high-angle boundary could be usefully analysed into dislocations itself has been questioned. The coincidence-site lattice model is considered to be physically significant only for those boundaries that have small angular ranges of misorientation (with reference to the coincidence-site lattice boundaries) (465,471). Aaron and Bolling (472) believe that the atomic packing at a high-angle boundary is so complex that there is no discernible structure. Pond and Smith (473), Ashby et al. (474) and Pond et al. (475), on the other hand, regard the grain boundary as an aggregate of 'random close packed' units (475) of atoms (analogous to the atomic arrangements in liquids (476)) interspersed with other non-random configurations (usually polyhedra) which are specific to the boundary. In a different field, Mott and Davis (477) have usefully compared the electrical reactance of grain boundaries in non-crystalline materials with that of a liquid layer.

Therefore, it seems more reasonable to regard grain boundary sliding to be the result of diffusion parallel to the boundary curvature at every point (63,357,381) until it is established that (a) dislocations are an integral part of the structure of high-angle boundaries, (b) they retain their integrity at the high homologous temperatures at which sliding is observed, and (c) they participate in sliding as dislocations, without dissociating into other types of defects.

In this approach the grain boundary is considered to be an aggregate of numerous, localized atomic scale units, flow within each of which is independent of others. Unit flow occurs by an atom-vacancy exchange and an accumulation of these atomic scale movements (at every point tangential to the local boundary curvature) leads to grain boundary sliding.

Both the unsteady (357,381) and the steady state (63,381) regions of optimal superplastic deformation have been attributed to "pure" grain boundary sliding. The width of the zone of deformation is essentially that of the grain boundary which, according to field-ion microscopic results, is two to three atoms wide. The less compact atomic packing encountered in this region ensures (a) poorer atomic fit and the presence of a greater number of vacancies necessary for the unit flow and (b) a lower activation energy for the atom-vacancy exchange parallel to the local boundary curvature.

In the initial, unsteady region atom-vacancy exchanges lead to grain boundary sliding until flow is blocked at a grain boundary obstacle, e.g. ledge, triple point. The resistance to flow offered by the obstacle leads to the development of an elastic back stress. When this stress significantly exceeds the mean boundary shear stress, stress-enhanced local diffusion which leads to atomic rearrangements will result. (It should be noted at this stage that in the earlier models (62,170) the back stress developed at the boundary obstacles was considered to be large enough to set up diffusion fluxes over distances of the order of (L/4). Thus, the minor rearrangements of the atoms (that are confined to the grain boundary region) required in this model constitute a (comparatively) much easier step.) This process will continue until the atomic arrangements in the vicinity of the (boundary) obstacles are made most conducive for easy and continuous sliding.

Once all the grain boundary obstacles are smoothed in this manner, steady state optimal superplastic flow (regions I and IIa of Fig.3.2(b)) could result merely from grain boundary sliding (which is defined above) because in this case the boundary obstacles are already removed and so flow will not be blocked (at triple points etc.). Then, macroscopic deformation is describable in terms of a three dimensional, non-Newtonian, continuously flowing network of grain boundaries, surrounding arrays of essentially non-deforming grains, in which the local velocity vector could possess any magnitude and direction due to the differences in the orientation and the nature of the interfaces as well as in the local atomic arrangements (within each boundary). If velocity discontinuities are to be avoided solely by the operation of this mechanism, the external rate of deformation should be consistent with the rate of diffusion along the grain boundaries (explains why superplasticity is lost above a critical strain rate) and the mobilities of the different types of interfaces present should be comparable (63,381) (explains two (equivalent) empirical observations: (a) intercrystalline and interphase boundaries are similar in their contribution to superplastic flow, although small differences could exist and (b) the mechanical properties of the phases should be similar under conditions of superplastic deformation (1,68)). Therefore, for the analysis the different types of boundaries have been replaced by an "average" boundary. (A procedure for taking into account the nature of the interface, when necessary, has also been outlined elsewhere (72).) Evidently, the real activation energy for deformation in both the unsteady and the steady state regions of flow will be of the order of the activation energy for grain boundary diffusion in the alloy.

However, as the deformation is confined to the grain boundary region and is diffusion-aided, significant contribution from this mechanism of flow will be present only in grain-growth resistant, ultra-fine grained materials (large grain boundary area per unit volume) that deform at high homologous ($T \gtrsim 0.4\ T_M$) temperatures (at which atomic mobility is high).

It should be noted that this mechanism has little resemblance to the other mechanisms of superplastic flow (Sections 4.2.3(a) - 4.2.3(c)) which invoke sliding as a rapid accommodating step. It is also quite different from the models of Stevens (396) and Cannon (397) for "diffusion creep" in which, regardless of the rate controlling step, the contribution from sliding does not exceed about 60-65%. To the best of our knowledge this is the first model which suggested that grain boundary sliding could be the rate controlling step and that under suitable experimental conditions sliding could give rise to extreme elongations without the aid of any accommodating process. A very early model due to Kê (478) also differs from this approach in a number of ways (63,381).

A number of experimental results are consistent with this alternative approach: (i) Experiments have revealed that "only portions of a boundary slide at any one time" (405). (ii) As a result of superplastic deformation grains acquire a bulbous shape (100,211,241) and this observation could mean that the triple points have been rounded and the ledges have been smoothed (63,357,381). (iii) Nicholson (68) has suggested that the experimental results are understood best if the entire (superplastic) deformation is attributed to grain boundary processes. (iv) Superplastically deformed alloys exhibit no strain hardening (1,68,129). (v) No experimental evidence for the blocking of sliding at the boundary obstacles during optimal superplastic flow has so far been reported. (vi) No evidence for the presence of absorbed dislocations at the grain boundaries of any superplastically deformed alloy is presently available (129). (vii) According to Harris (479) the temperature dependence of sliding is too strong to explain in terms of dislocation processes. (viii) It has been claimed (479,480) that some of the recent results concerning high temperature creep and diffusion in high-angle interphase interfaces are consistent with viscous intercrystalline flow models.

By postulating that the (average) time necessary to attain steady flow, t_s, will be of the order of the time necessary for an atom to cover a distance equal to the length of the grain boundary if the atom were to move down a potential gradient equal to that caused at the boundary by the action of the applied stress (as in this time all the obstacles between the two ends of the boundary would have undergone the necessary atomic rearrangements and evidently this smoothing process will go on simultaneously at all the boundaries), it has been shown that

$$(L/2) = \frac{D_{gb}\, a\, C}{kT} \int_{0}^{t_s} \tau \, dt \qquad .. \quad 4.38$$

where a is the area of cross-section of an atom, τ is the applied shear stress (which will be of the same order as the mean boundary shear stress), C is the concentration of vacancies and t is the time of deformation. (Evidently, the integral on the right hand side is equal to the area under the shear stress-time of deformation curve up to $t = t_s$.) Using the experimental data pertaining to the Sn-Pb eutectic alloy, the required values of D_{gb} for the model have been predicted (see Table 2 of ref. (357)). Gjostein (481), on the other hand, has represented diffusion data concerning non-FCC metals by the empirical equation

$$D_{gb} = 0.2 \exp(-20T_M/RT) \qquad .. \quad 4.39$$

where the gas constant, R, is expressed in calories mol^{-1}. The values of D_{gb} calculated from this equation (Table 2 of ref. (357)) are in good agreement with those predicted by the model. The experimentally measured value of the activation energy for grain boundary diffusion in tin (the faster diffusing species) (482) is also consistent with the predictions of the model.

The shape of the true stress-true strain curves in the unsteady region depended on the experimental variables (Fig.4.7). Both the true stress and the true strain at the onset of steady state flow increased with the rate of deformation but decreased with a rise in temperature. In the same material, depending upon the strain rate level, strain softening, an essentially strain-independent behaviour or strain hardening was observed. The minimum strain rate above which strain softening became perceptible increased with grain size. But strain softening was more pronounced on increasing the strain rate. If the temperature of deformation was raised (while keeping the strain rate and the grain size constant) the initial region of strain softening gradually changed to one exhibiting strain hardening.

In the model the above variations arise from the change in the relative importance of the following factors (357):

(i) In a constant velocity test with increasing strain the

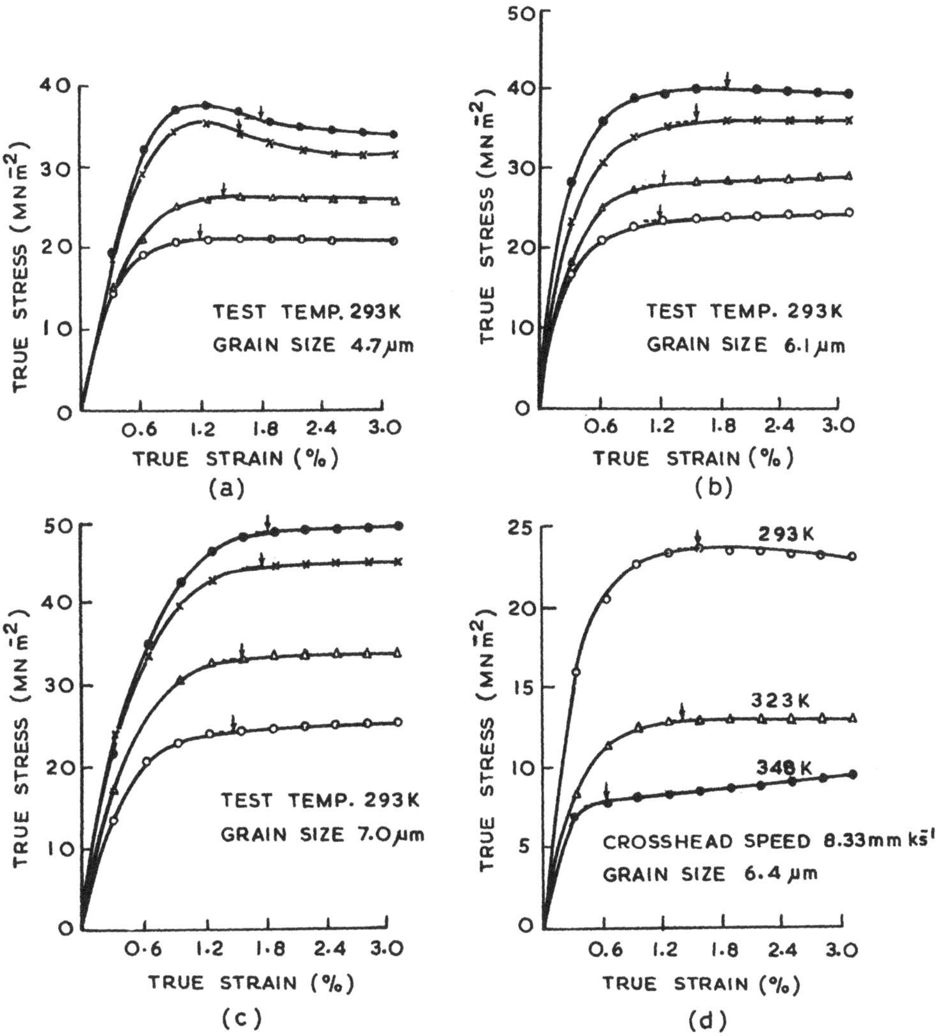

Figure 4.7 True stress-true strain relation in a Sn-Pb alloy in the region of transition from unsteady to steady state superplastic flow. (a)-(c) dependence on grain size (crosshead speeds: O 8.33 mm ks^{-1}, Δ 16.6 mm ks^{-1}, X 83.3 mm ks^{-1}, □ 166.6 mm ks^{-1}), (d) as a function of temperature; the onset of steady state indicated by arrows (Yadava and Padmanabhan (357))

strain rate and hence the flow stress decreases. (ii) Grain growth becomes more significant with an increase in temperature or a decrease in strain rate. (A direct relationship exists between the flow stress and the grain size.) (iii) In the experiments m increased with grain size. Therefore, at larger grain sizes the effect of the decrease in strain rate with deformation (point(i)) became less important. (iv) In view of the shorter time available, the excess stress (above the steady state value) necessary to smooth the boundaries increases with an increase in strain rate. The shorter time also ensures smaller grain growth. Evidently, both these factors lead to more pronounced strain softening (with increasing strain rate). (v) On account of the direct relationship between the steady state flow stress and the grain size, the excess stress necessary to smooth the boundaries decreases with increasing grain size (if the grain boundary roughness does not change with grain size in the experimental range). (vi) When the strain rates are greater than that corresponding to maximum m (in which range dislocation motion is present), a rearrangement of dislocations (as in hot working) could also contribute to strain softening.

Steady state flow, on the other hand, is described by the equation

$$\dot{\gamma} = c\ N\ \gamma\ \nu \left[\exp\left(\frac{-Q+f(\tau)}{kT}\right) - \exp\left(\frac{-Q-f(\tau)}{kT}\right)\right] \qquad .. \quad 4.40$$

where $\dot{\gamma}$ is the macroscopic shear strain rate, N is the number of active sites at which a unit jump is occurring, γ is the mean value of the localised shear strain associated with each jump, ν $(=(kT/h))$ is the frequency of thermal vibration, h is Planck's constant, τ is the mean boundary shear stress and c is a constant of the order of unity which connects the local shear rate with the macroscopic shear rate.

N will depend directly on the total grain boundary volume available. Then,

$$N \propto a_1\ w\ g$$

where a_1, the surface area of a grain, varies as L^2, w the width of the grain boundary is essentially of constant value (as indicated by field-ion microscopy) and g the number of grains per unit volume will be equal to (c_3/L^3), where the dimensionless constant c_3 takes care of the actual shape of the grains. Therefore, $N = (c_4/L)$ where c_4 (which has absorbed w) has the unit of length.

From the definition of shear strain, a unit jump produces a strain of (d/r) where d, the jump distance, will be equal to the interatomic spacing in the grain boundary region and r is the equivalent radius of the grain (as the shape will not be exactly spherical). As this strain is shared by two grains, $\gamma = (d/L)$.

$f(\tau)$, the work done by the applied stress, is given by $\beta\tau^{*n}$, where β is a constant that has the dimensions of energy, $\tau^{*} = (\tau/e\tau_c)$ is a non-dimensional stress, e is the base of the natural logarithms and τ_c is the critical stress for which m=1. The simplest solution for the stress sensitivity, n (=1/m), is of the type

$$n = \left(a_1 + \frac{a_2}{\ln\tau^*}\right)\left(b_1 + \frac{b_2}{T}\right)(c_1 + c_2 \ln L) \qquad .. \quad 4.41$$

where a_1, a_2, b_1, b_2, c_1, c_2 are constants and n is always greater than or equal to unity. (The procedure for obtaining more complicated solutions has also been outlined. Moreover, the functional given by equation 4.41 has both theoretical and experimental justification (63).)

As $\beta\tau^{*n} \ll kT$ and $\nu = (kT/h)$, equation 4.40 becomes

$$\dot{\gamma} = c_1^1 \frac{d}{L^2} \tau^{*n} \exp(-Q/kT) \qquad .. \quad 4.42$$

if $c_1^1 \equiv (2\,c\,c_4\,\beta/h)$. (As h has the dimensions of (energy. time), c_1^1 will have the units of (length. time^{-1}).)

For metallic systems d is approximately constant and is typically of the order of 2-3 Å. Moreover, during plastic flow a direct relationship between the shear and the corresponding tensile quantities exists. Therefore,

$$\dot{\epsilon}_t \propto (\sigma_t^{1/m}/L^2) \exp(-Q/kT) \qquad .. \quad 4.43$$

which is analogous to the empirical equation 3.9. During isothermal deformation over narrow ranges of strain rate m is relatively constant and then equation 4.43 will reduce to the popular, empirical equation 1.2.

The real activation energy, Q, can be evaluated only when the variations in m(=1/n) with the experimental variables are taken into account. The procedure (69,319,323,334) has already been outlined in

Section 4.2.2. Briefly, both the (apparent) activation energy determined at constant stress, Q_{τ} , and the (apparent) activation energy evaluated at constant strain rate, $Q_{\dot{\gamma}}$, vary widely (and their magnitudes depend upon the local value of m). The interdependence between Q and Q_{τ} is given by equation 4.14 while that between Q and $Q_{\dot{\gamma}}$ is governed by equation 4.15. When m is relatively constant, Q_{τ} is related to $Q_{\dot{\gamma}}$ by equation 4.12.

From equations 4.14 and 4.41 it follows that Q_{τ} decreases parabolically with increasing magnitude of the (constant) applied stress (381). Then in regions where n is high (at low stresses and temperatures), Q_{τ} could be mistakenly identified as the activation energy for volume diffusion.

As the inequalities $Q_{\dot{\gamma}} < Q < Q_{\tau}$ and $(Q_{\tau} - Q) > (Q - Q_{\dot{\gamma}})$ are valid (Section 4.2.2), the empirical observation (69,323) that $Q_{\dot{\gamma}}$ is closer to Q compared with Q_{τ} can be accounted for. The grain size dependence of the apparent activation energy (section 4.2.2), on the other hand, is traced to the grain size dependence of n (equation 4.41). Finally, as $m \to 1$, the values of both Q_{τ} and $Q_{\dot{\gamma}}$ tend to that of Q.

Climb-controlled motion of dislocations is considered to be the dominant mechanism in the region beyond peak m (regions IIb and III in Fig.3.2(b)) where superplasticity is gradually lost. For describing flow in this region either equation 4.24 (Ashby and Verrall (62)) or equation 4.36 (Edington et al. (129)) is invoked. However, both these equations are empirical. Therefore, although the point at which superplastic tendency starts decreasing could be considered to coincide with the onset of dislocation motion, the flow stress at this stage (as a function of grain size, strain rate and temperature) is yet to be evaluated theoretically.

Evidently, there will be an intermediate region (beyond the point of commencement of dislocation motion) in which both the viscous boundary flow process and crystallographic slip will coexist. A crude way of predicting the overall behaviour in this region is to add equation 4.42 with the appropriate expression corresponding to region III of Fig.3.2(a) (see above). However, as both these processes will compete for vacancies, strictly the two components should not be added. Then, the domain in which both the mechanisms will coexist will be narrower than what is predicted by simple addition.

To summarise, at low stresses up to $\tau \simeq \tau_c$ the viscous flow mechanism operates. In some systems, however, the stress necessary to start dislocation motion could be less than τ_c. Then, the decrease in superplastic tendency will commence at the former stress. Evidently, for these systems the maximum value of m will be less than unity (and is obtained from equation 4.41 by substituting the value of the stress at which dislocation motion commences). Once dislocation activity starts, there will be a region in which both viscous boundary flow and dislocation motion will be simultaneously present. Beyond this range climb-controlled motion of dislocations will be predominant.

The above analysis has been shown to predict the correct strain rates in case of the Al-Cu eutectic and the Zn-Al eutectoid alloys and a Ni-Cr-Fe alloy. The predicted stress and temperature dependence of n (equation 4.41) also has experimental support. Although the predicted logarithmic variation of n with grain size appears to be realistic, paucity of experimental data has prevented actual verification.

The approach has the following additional implications:

(i) Grain elongation and the ease of solute redistribution (to avoid excessively large diffusion paths) required by other diffusion-based models are dispensed with. Instead, intergranular displacements and grain rotation could occur in much the same way as during the flow of viscous suspensions. Moreover, Beere (386,387) has shown that during grain boundary sliding controlled creep grain rotation and grain switching are automatically predicted, if the condition of coherence of deformation is imposed. Beere's analysis also demonstrates that the problems of maintaining the coherence of deformation and the force equilibrium as well as starting the postulated atomic fluxes are absent in this model (63,357,381), unlike in case of the mechanism proposed by Ashby and Verrall (416). As seen earlier, in general diffusion parallel to the boundary is easier than along the normal direction. However, if in a local region of the boundary the atomic arrangement is complex or alternatively if the chemical potential of the vacancies on either side of a boundary (e.g. an interphase interface) is different, diffusion in a perpendicular direction becomes equally likely (particularly when the boundary shear stress is small). Thus, this approach can account for the occasional grain elongation seen in region I (Fig. 3.2(a)). (It should be noted that the results of the analysis of sliding controlled creep are unaltered even if the normal stresses are not fully relaxed (387).) The experimental procedure of attributing the normal displacements to diffusion and the parallel components to sliding to conclude that the sliding contribution is more in region II than in region I, however, becomes meaningless because in this approach sliding is also caused by diffusion.

(ii) In two phase materials diffusive displacements could occur predominantly along the more mobile boundaries. Then the interconnection of the less mobile phase should decrease in the direction of deformation. Spheroidization of the less deformable phase should also be present. (Both these predictions have been experimentally verified (241).)

(iii) The lack of evidence for dislocation activity during superplastic flow (Section 3.2) is consistent with this model. In structurally uniform material, as is only to be expected, the geometry of cross section is preserved after extreme elongation (241). Thus in this approach the elliptical cross section developed in hot rolled Zn-Al eutectic alloy specimens as a result of superplastic flow (194) is traced to the non-uniform initial microstructure.

(iv) The interpretation of Edington and coworkers (see, for example, ref. (129)) that the results of the marker experiments (Section 3.2.1(c)) and the texture studies (Section 3.2.2(c)) are consistent with the presence of dislocation activity, on the other hand, has already been shown to be equivocal.

In the present approach the retention of some of the texture components following superplastic flow is due to the differences in the sliding resistance of the different types of boundaries (A-A, A-B or B-B, where A and B are the component phases), their orientation with respect to the stress axis and the minor variations in the atomic arrangements at each of these boundaries. Sliding on the more viscous type of boundaries can be reduced by grain rotation (387). When the most viscous type of boundaries are perpendicular to the direction of stress, little sliding will take place (on this type of interface).

(v) In region I significant flow should occur only at those

boundaries where the resolved shear stress is high. Then, the deformation will be localised and superplasticity is less significant. Equation 4.41 also predicts this result because m is small at very low stresses (Sections 2.5 and 2.6).

(vi) This approach can explain with the aid of a single mechanism the observation that superplastic characteristics are enhanced with an increase in stress to a maximum before climb-controlled dislocation motion gradually eliminates superplastic behaviour. (No use is made of the concept of a 'back' or 'threshold' stress.)

(vii) When $n \rightarrow 1$ (equation 4.41) the present formulation for $f(\tau)$ reduces to the earlier (linear) representations (382,383,442,478, 483). Then, the experimental observation that Newtonian viscous flow (n=1) occurs over a small range of stress at very high temperatures could be justified as a special case of this more general analysis.

(viii) The mechanism of flow is isotropic, particularly in materials made up of random, equiaxed grains. The virtual absence of elastic strains and the low magnitude of the hydrostatic component of stress, on the other hand, ensure that the alloy is incompressible during optimal superplastic deformation. Both these predictions have experimental support (99,484).

The four criticisms of the model by Gifkins and Langdon (439), on the other hand, have been refuted (381) on the following grounds:

(i) The suggestion (439) that sliding accommodated by either diffusion or dislocation motion could also explain superplastic flow is not consistent with the analysis of Beere (386,387) which shows that significant grain rotation can result only when sliding is the rate controlling step.

(ii) The conclusion (439) that the thickness of the deforming "mantle" will be considerably more than the grain boundary width if flow is confined to the grain boundary region is not valid because the calculations are based on the "core and mantle" mechanism of Gifkins (388) (and not on the model of Padmanabhan (63,357,381),.

(iii) The suggestions (439) that in equation 4.42 (a) the inverse square dependence (of strain rate) on grain size that has been deduced is dubious and (b) the physical significance and the dimensional requirements of c_1^1 have not been explained, originate from a partial understanding of the analytical procedure (381). (See above for detailed derivation.)

(iv) The suggestion that Padmanabhan has attributed the variations in the magnitude of the activation energy (Q_τ, which has been taken by Gifkins and Langdon (439) to be the real activation energy) to a lack of distinction between Q_τ and $Q_{\dot{\gamma}}$ is the result of an inadequate survey of the available literature. As has already been pointed out, in this approach the variations in Q_τ should be traced to equation 4.14.

In conclusion it could be said that (for reasons stated earlier) the models of Gifkins (388) and Gittus (389) are inadequate. The mechanism proposed by Padmanabhan (63,357,381) is able to account for most of the unequivocal as well as the conflicting experimental results concerning superplastic flow. It also predicts the strain rates and the interdependence of variables correctly. The phenomenological analysis of Beere (386,387) is consistent with the atomistic model of Padmanabhan.

However, the general acceptance of the theory of Padmanabhan (63,357,381) will depend to a large extent on the acceptance of two recent ideas: (a) perhaps much point is not served by analysing high-angle grain boundaries into dislocations if the aim is to understand their equilibrium and non-equilibrium behaviour and properties (474) and (b) "pure" grain boundary sliding can lead to extreme elongations without the aid of any accommodating process (63), once the boundary obstacles are removed by the mechanism outlined earlier (357). Therefore, at present it is safe to regard this theory as a useful mathematical construct which is able to account for most of the experimental observations concerning superplasticity.

(d) Multi-mechanisms

In some cases superplastic flow has been considered to arise from the simultaneous operation of a number of mechanisms. The models (in this category) in which the rate controlling step is unique have already been reviewed in one of the sub-sections 4.2.3(a)-4.2.3(c). Occasionally, more than one process contributes to the overall strain rate. Some of these (the phenomenological approaches) were presented in Section 4.2.1 and the rest (the models that invoke atomistic mechanisms) are considered below.

Following Weertman and Weertman (485), Mohamed and Langdon (486) have suggested that deformation mechanism maps (originally called "creep diagrams") could be constructed (in case of superplastic alloys) by taking the applied stress and the grain size as the independent variables. In these maps the boundaries between the different mechanisms are straight lines, unlike in the earlier constructs (487) where the independent variables were the stress and the temperature of deformation. Obviously, these maps can also be constructed by using the grain size and the (homologous) temperature of deformation as the independent variables (488). No doubt these deformation mechanism maps represent a simple and highly visual method of describing the mechanical behaviour of materials. However, the technique, per se, can provide no insight into the nature of the flow processes.

According to Dawson (317) strain-rate sensitivity of materials arises from the competing effects of strain hardening and time dependent thermal softening. Therefore, the flow stress is a function of the current value of strain and the time of deformation. (Evidently, the usefulness of this type of analysis in understanding superplastic flow is rather limited - Sections 3.1 and 3.2.) The technique appears to be little more than an (empirical) curve fitting exercise, because the total stress has been divided somewhat arbitrarily into two parts (to allow for the above two mechanisms). Moreover, no reduction in the number of independent variables is achieved when the strain rate gets replaced (in the analysis) by the lapse time.

A number of speculations are available as to the origin of superplastic flow: grain boundary sliding, diffusion and boundary migration (489); grain boundary shear and recrystallisation (100,490); grain boundary shear, Nabarro-Herring diffusion, recrystallisation and/or recovery (491); grain boundary sliding and crystallographic slip (214); and sliding, slip, diffusion and migration (205,310,361, 492) are some of the combinations suggested.

Lee (230) regards crystallographic slip as the rate controlling mechanism at the highest strain rates. With decreasing strain rate sliding becomes increasingly important. Grain deformation and recovery ("slip-recovery cycle") ensure continual sliding. Grain rotation and boundary migration could also be important. The dominant mechanism at the lowest strain rates is either diffusional flow or grain boundary sliding. This model is also qualitative.

Avery and Backofen (70) have proposed the equation

$$\dot{\epsilon}_t = A' \rho D_1 \ \text{Sinh} \ (v^* \sigma_t / kT) + (\alpha \Omega D_1 \sigma_t / L^2 \ kT) \quad .. \quad 4.44$$

where A' is a constant that includes the Burgers vector and the inter-jog distance, ρ is the mobile dislocation density, v* is the activation volume and α is approximately equal to 10, to describe superplastic flow. The first term is due to dislocation motion while the second term represents Nabarro-Herring diffusional flow. At very low strain rates the second term is considered to be dominant and m=1. With increasing strain rate the first term becomes increasingly important and m continuously decreases. On account of the second term, with a decrease in grain size superplasticity persists to higher strain rates. Backofen et al. (42), however, had observed that m goes through a maximum with strain rate (and this is the more general result (1,68, 129,326)). Although this could be interpreted in terms of a strain-rate dependent diffusivity, no experimental support for this concept could be found.

Packer and Sherby (316), however, have concluded that (a) in two phase alloys contribution from Nabarro-Herring diffusional flow is negligible, (b) at very low strain rates m is not equal to unity and (c) because of the very nature of the expression at all strain rates the first term of equation 4.44 should dominate. So they have proposed an alternative equation

$$\dot{\epsilon}_t = B \sigma_t^2 \ \text{Sinh} \ (\beta \sigma_t^{2.5}) + \frac{A \sigma_t^2}{L^3} \quad .. \quad 4.45$$

where B, β and A are constants. In this case also the first term is due to dislocation creep but the second term represents continuous recrystallisation or boundary migration. If the constants were evaluated using earlier data (70), poor correlation with the new results (316) was found. Moreover, the concepts of continuous recrystallisation and migration have little relevance to superplastic flow (Sections 4.2.3(a) and 4.2.3(b)).

An alternative expression due to Packer (493) is given by

$$\dot{\epsilon}_t = B \sinh \ (\beta \sigma_t) + \frac{A\sigma_t^2}{L^3} (\exp (-\eta/\sigma_t)) \qquad .. \ 4.46$$

in which the first term is due to dislocation creep and the second term arises from recrystallisation. The constant η has been introduced to predict a relatively low degree of recrystallisation at low stresses. Then, at both low and high stresses dislocations creep will dominate. It is not clear in what way the exponential term reflects the fraction of grains that recrystallise. Moreover, the form of the expression assumed for dislocation motion predicts m of the order of unity at very low stresses. The second term, on the other hand, has its origin in the controversial concept of dynamic recrystallisation (Section 4.2.3 (b)).

Baudelet and coworkers (82,144,304) have considered the superplastic deformation of microduplex materials in which initially the grains of the less mobile phase are elongated. During deformation these grains slide along the interphase interfaces and align themselves parallel to the tensile direction. Subsequently, the elongated grains separate along their internal grain boundaries and achieve a nearly equiaxed shape. This step requires a higher stress compared with steady state deformation and so there is an initial region of strain softening (304). As a result, steady state flow commences only after an elongation of about 70% (144).

Steady state superplastic deformation, on the other hand, has been attributed to the combined action of grain boundary sliding and deformation of the softer phase, with the latter controlling the overall rate of flow. An expression

$$\dot{\epsilon}_t \simeq \frac{A'\sigma_t^2}{\bar{L}\, L_\alpha^2} \exp (-Q/RT) \qquad .. \ 4.47$$

with A' a constant, $\bar{L}$ the average grain size, L_α the grain size of the more ductile phase and Q the activation energy for volume diffusion (when the microstructural changes accompanying the deformation are not ignored) has been obtained. It has further been suggested that when the structure is stable, m is independent of both strain and strain rate and is equal to 0.5.

The physical picture (on which the analysis is based) is unclear. The intrinsic rate of the accommodating mechanism (deformation of the softer phase) has been assumed to be equal to $A\,\sigma_t^2\, L_\alpha^{-3}\, \exp(-Q/RT)$. If the accommodation is achieved by diffusion, a stress sensitivity of 2 is unlikely and the inverse cube dependence (of strain rate) on grain size would suggest that the activation energy should be of the order of that for grain boundary diffusion (Coble creep). On the other hand, if dislocation motion accommodates sliding an inverse or inverse square (but not inverse cube) dependence of strain rate on grain size can be obtained only when the grain boundaries act as the sources of dislocations (Section 4.2.3(b)). Then, Q should be equal to that for grain boundary diffusion. If bulk diffusion is sufficiently rapid m will be of the order of 0.33 and the strain rate will be independent of grain size. As regards the value of m, it has been shown in Section 3.1.3 that even when structural changes are negligible, e.g. as in a stress relaxation or a change in strain rate test, m could be significantly different from 0.5.

While assessing the usefulness of the multi-mechanisms Nicholson (68,291) has noted that the dislocation models cannot predict $m > 0.5$ and for diffusional flow m=1. A suitable combination can give the required range of values for m but these mechanisms predict (contrary to experimental results) grain elongation. Continuous recrystallisation or dynamic recovery involving the continual migration of grain boundaries, on the other hand, can ensure the retention of an equiaxed grain shape but these processes cannot be invoked in case of systems in which the phases differ widely in composition, e.g. the Cd-Bi eutectic alloy. Moreover, models based on recovery and recrystallisation are extremely unrealistic because cyclic hardening and softening, softening on unloading and the presence of a well-defined yield point (which are predicted by these models) are seldom observed during superplastic flow (1,315). Finally, it is emphasized that on account of the available flexibility numerical agreement with data can easily be obtained if multi-mechanisms are invoked. Therefore, experimental verification (for the presence) of each mechanism is essential.

5. Environmental Superplasticity

In Section 1.2 the discovery of transformational plasticity and the early work in this area up to late 1950's was described. As briefly noted in that section environmental superplasticity is caused by any one of the three methods:

(i) thermal cycling of a material through a phase change with concurrent application of a small external stress,

(ii) thermally cycling a material with anisotropic coefficients of thermal expansion while simultaneously applying a load, and

(iii) subjecting a material under a small stress to neutron irradiation.

For optimal effects thermal cycling in the first two cases is done repeatedly when more than 500% elongation could be obtained. Large elongations, however, have not been produced by neutron irradiation because of the extremely low rates of deformation. Edington et al. (129) compiled in 1976 a list of materials (Appendix 2 of ref. (129)) in which environmental superplasticity has been established. There has virtually been no addition to this list since then. Moreover, from this list it will be obvious that most of the studies have been concerned with transformational plasticity.

5.1 Experimental Results

5.1.1 Macroscopic characteristics

(a) Transformational plasticity

In the early literature transformational plasticity was also referred to as "amorphous plasticity" (see, for example, Clinard and Sherby (494)). This phenomenon has been established in a number of ferrous and non-ferrous materials. Both diffusion-controlled as well as diffusionless phase transformations have been made use of. Therefore, it is possible to identify a few general characteristics of transformational plasticity.

As in the case of structural superplasticity, transformational superplasticity also gives rise to large elongations at very low applied stresses. Grain size, however, has negligible importance. As martensitic transformations occur at very low temperatures perhaps there is no restriction on the temperature of occurrence of transformational plasticity. Studies involving the use of tensile, torsional or compressive forces reveal that the mode of application of stress is unimportant. In many cases the geometry and the purity of specimens used had little influence (29,148-150,494-497).

Fig.5.1, due to de Jong and Rathenau (27), indicates the

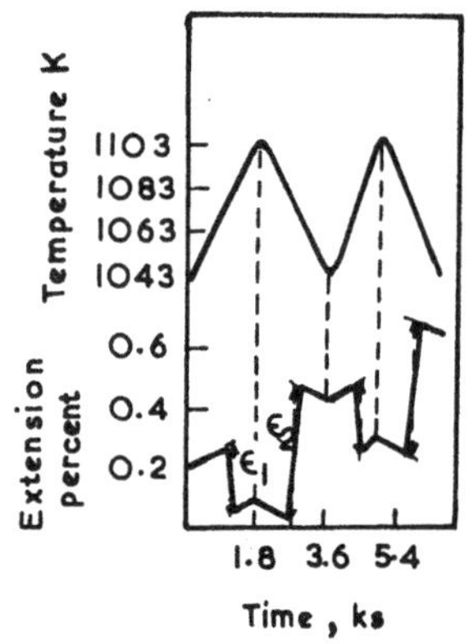

Figure 5.1 Elongation of a polycrystalline iron specimen under a small load during the $\alpha \rightarrow \gamma$ and the reverse transformation (de Jong and Rathenau (27))

dimensional and the temperature changes produced in a polycrystalline iron sample when it is subjected under a small stress to a thermal cycle of 3.6 ks (60 minutes) duration through a range that includes the $\alpha \rightleftharpoons \gamma$ transformation. The excess strain arising from the $\alpha \rightarrow \gamma$ transformation is ϵ_1 while the strain due to the $\gamma \rightarrow \alpha$ transformation is ϵ_2. Fig.5.2, which is derived from figures similar to Fig.5.1, presents ϵ_1, ϵ_2 and ($\epsilon_1 + \epsilon_2$) as functions of the applied stress. It should be noted that the plot between ($\epsilon_1 + \epsilon_2$) and the applied stress passes through the origin. Moreover, the curves for ϵ_1 and ϵ_2 are parallel and thus deformation in this case is independent of the direction of phase transition. This is also the case with commercial purity titanium and the Fe-32.9% Ni alloy (148,498). In case of many steels

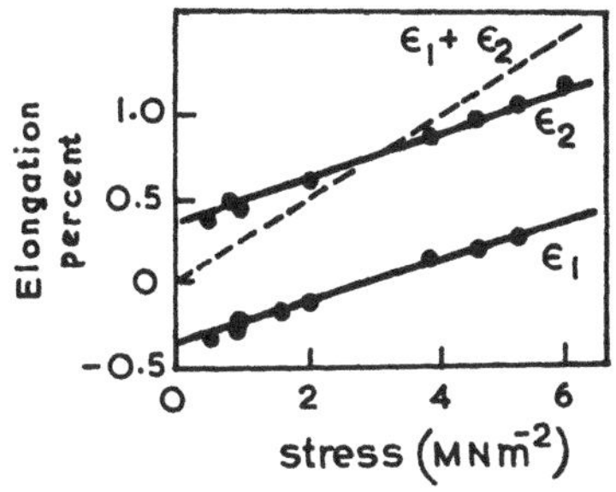

Figure 5.2 The changes in length of a polycrystalline Fe-0.008 wt.% N specimen as a function of applied stress during the $\alpha \rightleftharpoons \gamma$ two-way transformation. ($\epsilon_1 + \epsilon_2$) is the strain per cycle (dashed line). (de Jong and Rathenau (27))

and the Fe-15.4% Ni alloy, however, deformation during heating is greater than that during cooling (29,148,494,498). There is also a case where the converse is true (496).

In one case (27) where the boundary between the parent and the product phases was planar, elongation was independent of applied stress. In general, however, a direct linear relationship exists between the excess strain due to transformation and the applied stress, so long as the latter is below a critical value (27,498-500).

Excess elongation due to transformational plasticity is a linear function of the fractional change in specific volume (=($\Delta V/V$)) accompanying the phase transition. It also depends on the mechanical strength of the weaker phase. Internal stresses arising from phase transformation are usually of the order of the yield stress of this phase. Then deformation could occur at very small external stresses. Naturally relaxation will be confined to the weaker phase (15J,501).

Excess elongation increased with the amplitude (= $T_{maximum} - T_{minimum} = \Delta T$) of the thermal cycle. Moreover, it was necessary that ΔT had a minimum value to be able to produce any excess elongation at all (27). Elongation at constant stress, on the other hand, varied linearly with the amount of material transformed in a given reaction (27,499,500,502). In one case the mean temperature of the cycle, T_M, had no effect on elongation (27) but in another the elongation was maximum when the mean temperature coincided with that of the phase transition (500). When ΔT was large complete transformation occurred in each cycle and then deformation per cycle was independent of the rates of heating and cooling. On the other hand, when ΔT was very small, the strain per cycle was constant and independent of the number of cycles employed (500). However, according to an earlier study (494) the strain per cycle decreased with increasing rates of heating and cooling. Holding the specimen at the extremities of the temperature cycle decreased the magnitude of deformation per cycle. When the holding time was very large the excess elongation was close to zero (498). As transformational plasticity was affected by the ($\Delta V/V$) ratio, strain per cycle (at constant stress) produced in steels decreased with increasing carbon content (503). Figs.5.3(a) and (b), due to Petsche and Stangler (503), are three dimensional plots which show the dependence of deformation per cycle on T_M and the carbon content.

For many steels the deformation per cycle - applied stress curves were parallel. But when specimens were temperature cycled through a phase change without the application of an external stress, the sign and the amount of net deformation that resulted depended on composition. In plain carbon steels, for example, the net change in length was zero when the carbon content was 0.4%. When the carbon level was higher there was elongation. Contraction was observed when the carbon content was less than 0.4% (29,498,499). In the latter case, therefore, a small positive stress has to be applied before any excess elongation can result. This is occasionally referred to as the threshold stress necessary for inducing transformational plasticity (see, for example, Hoke et al. (499)).

In Armco iron (C=0.03%) when a temperature gradient existed along the axis of a specimen (with the maximum temperature being present at the middle of the gauge length), two necks which were symmetrically located about the centre formed. The temperature at the necked regions was in the range 993-1123 K. The distance between the necks increased with an increase in the magnitude of the temperature gradient. The deformation at the necks was considered as evidence for the occurrence of "superhigh plasticity" resulting from the transformation along the grain boundaries (28).

At low stresses at which the deformation per cycle increased linearly with the magnitude of the former no hardening or softening mechanisms were present (150). Fig.5.4 presents the strain per com-

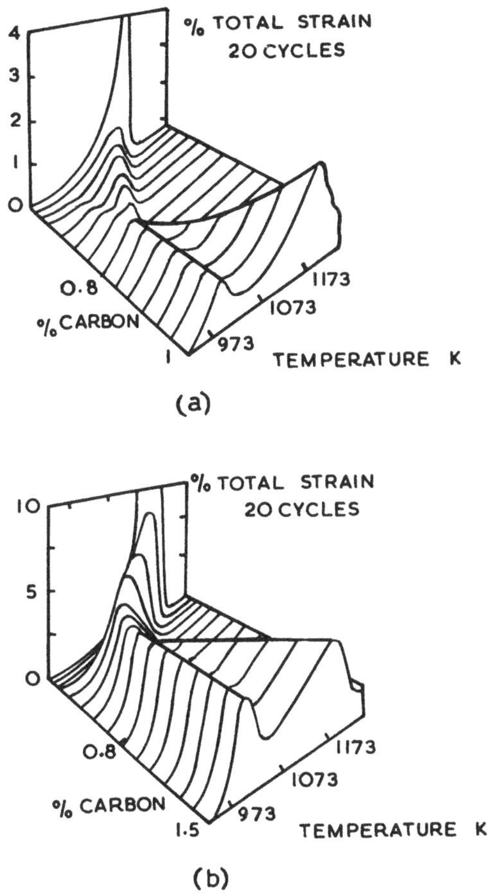

Figure 5.3 Accumulated compressive strain (after 20 temperature cycles) as a function of T_M and carbon content (applied stress = 5.7 MNm^{-2}).
(a) $\Delta T=30^{o}C$; heating rate = 166.7^{o}C ks^{-1} (10°C min^{-1})
(b) $\Delta T=125^{o}C$; heating rate = 666.8^{o}C ks^{-1} (40°C min^{-1})
(Petsche and Stangler (503))

plete two way cycle as a function of the applied stress for the $\beta \rightleftharpoons \gamma$ transformation in uranium at 1043K. It is evident that at high stresses the relationship becomes non-linear. Commenting on this result Greenwood and Johnson (150) have remarked that the non-linearity is the result of significant continuing deformation of the weaker phase under the applied stress in the absence of transformational stresses. Normal

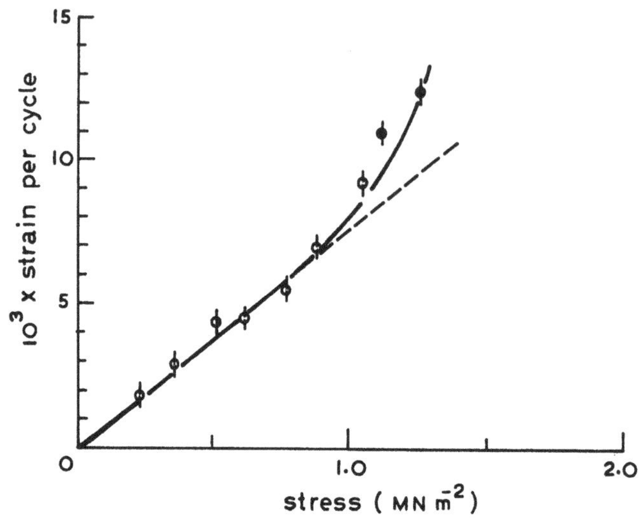

Figure 5.4 Strain per complete two-way cycle plotted as a function of applied stress for the $\beta \rightleftharpoons \gamma$ transformation in uranium at 1043K (Greenwood and Johnson (150))

plastic deformation often exceeded that resulting from transformation and eventually fracture occurred but only after the material had transformed to the weaker phase.

Transformational plasticity, based on the martensitic transformation, has also been studied in a number of alloys (504-508). In some cases a threshold stress had to be applied before excess elongation could be produced during transformation (507,509). As in the case of diffusion-controlled transformations, excess deformation during martensitic transformation also increased linearly with applied stress. Moreover, it was also reciprocally related to the yield stress of the weaker phase. Alloy additions which either suppressed the transformation, e.g. addition of vanadium to an Fe-Mn alloy, or brought about precipitation hardening eliminated transformational plasticity. An increase in the austenitizing time and temperature, on the other hand, increased transformational plasticity perhaps by raising the M_s temperature (506,507,509). In some cases the M_s temperature was raised by the application of stress (505,507) but in others the applied stress had no effect on the M_s temperature or on the amount of martensite formed until very high stresses were reached (510). In sheets of

Fe-30% Ni alloy having cubic texture transformational plasticity depended also on the orientation between the tensile and the rolling directions. The orientation effect was maximum when the angle between the two axes was 22.5^{o}. As with transformational plasticity induced by diffusion-controlled transformation, at large stresses the linear dependence of strain on stress vanished (505,509). During the reverse transformation from martensite to the parent phase in many cases due to mechanical hysterisis only a portion of the transformational strain was recovered. The magnitude of the reversible component of strain, which decreased with increasing (applied) stress, was also dependent on the orientation between the tensile and the rolling directions. An intriguing observation, however, was that under macroscopically identical conditions the transformational plasticity obtained varied significantly, over and above what could be attributed to experimental errors (509).

Occasionally during transformation softening arising from pearlitic, bainitic or martensitic reactions the applied stress produced very rapid strains in the early stages of transformation. In the later part, however, the transformational strain was time-dependent (129,510). It has also been noted that the dependence of transformational plasticity on the experimental variables during tempering of martensite or an isothermal transformation was analogous to the results described earlier (511).

In some cases excess deformation could be obtained by the use of order-disorder transformation (Ti-Ni system), precipitation reaction (Sn-Sb and $Zr-H_2$ systems), recrystallization or controlled martensitic transformation (metastable austenitic stainless steels) (29,129,512-514). In high strength steels, on the other hand, 'transformation induced plasticity' (TRIP) can produce elongations of the order of 100% (326). It is not, however, very clear whether these could be looked upon as examples of environmental superplasticity.

A phenomenon analogous to transformational plasticity, but impermanent in nature, is "superelasticity". It is observed when stress-induced (thermo-elastic) martensitic transformation occurs in a temperature range where the martensite formed is unstable in the absence of an external stress. Here the strain is completely recovered on unloading. The phenomenon is encountered in copper-base alloys, In-Tl alloys and the Au-Cd and Ti-Ni systems (515).

In case of ceramics the "Hedvall effect" (enhanced reactivity of a solid during phase transformation) has been used to obtain very dense and strong products by applying pressure during phase transformation. Then the temperature and time of pressure sintering could be significantly lowered. Following this method enhanced compaction has been achieved during the transformation of quartz to cristobalite, metakaolin to a spinel, monoclinic to tetragonal form of zirconia and the γ to α form of alumina. Enhanced creep rate, on the other hand, was obtained in quartz when the deformation was carried out during the $\alpha \rightarrow \beta$ transformation (516-519). In Bi_2O_3 excess elongation was observed (only) during the heating part of the temperature cycle (129). These phenomena encountered in ceramics have only a tenuous connection with environmental superplasticity because no unusual deformation at relatively low stresses has so far been reported.

Efforts to produce large specimen elongation using transformational plasticity are of relatively recent origin (29,148,498). Fig. 5.5 presents data pertaining to three plain carbon steels and a bearing

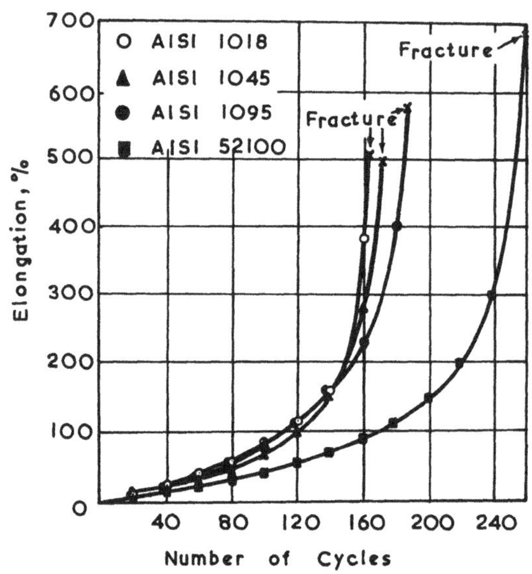

Figure 5.5 Deformation due to temperature cycling between 811K (1000°F) and 1088K (1500°F) versus number of cycles for four steels. Constant load tests, initial stress 17.1 MNm^{-2} (Oelschlaegel and Weiss (29))

steel. Deformation due to temperature cycling has been plotted as a function of the number of cycles applied. Elongation at fracture varied between 500 and 720%. In commercial grade titanium and the Fe-15.4% Ni alloy, on the other hand, elongation at fracture was 300 and 160% respectively (148,498). All these tests were performed at constant load where due to a decrease in cross-section elongation per cycle increased very rapidly in the last stages of deformation and led to rapid failure. It is, therefore, expected that tests at constant stress will produce even greater elongation (148).

There is a set of results (151) concerning a steel of eutectoid composition in which an elongation of about 490% was obtained by straining the material in the austenitic range (making use of the high strain-hardening capacity of austenite). Thermal cycling was used only to recover the initial flow stress by phase transformation. Care was also taken to see that the strain applied before each temperature cycle did not exceed the necking strain (Section 2.5.1). The elongation obtained in this case does not appear to be an example of environmental superplasticity but is analogous to the strain-recrystallization cycles employed by Lainer et al. (520) for obtaining large elongations. The only difference is that in the former case a phase change has been used in place of recrystallization to recover the flow stress.

Section 2.8 explains the procedure by which the strain-rate sensitivity index for environmentally superplastic flow is calculated to be of the order of unity. It was further pointed out that in view of the gross assumptions made in the analysis attempts at providing a physical basis for m will not be very meaningful (see also Edington et al. (129)).

Yoder and Weiss (151) have outlined an alternative method using steel specimens of eutectoid composition. The specimen is initially strained at 989K until a maximum in the applied load is obtained. Then the temperature is raised to that of the eutectoid transformation, i.e. 1000K. Now the material is strained at this temperature and m is calculated in the vicinity of the load maximum using the change in strain rate method (Section 2.7.2). In this case a value of 0.35 was obtained for m. The eutectoid transformation is known to be sluggish. Thus it is not clear how much transformation took place during the short time in which m was evaluated. Taken together with the fact that the m value obtained here was equal to that of the untransformed two phase material it is safe to conclude that this method also is not very useful in providing physical meaning for the strain-rate sensitivity index, particularly because many of the phase changes responsible for transformational plasticity take place rapidly.

(b) Temperature cycling

Most of the studies concerning the behaviour of materials with anisotropic coefficients of thermal expansion during temperature cycling have been confined to high purity and fuel grade "adjusted" (total impurity content $\sim$2000 ppm; chiefly C, Al, Fe and Si) uranium and high purity zinc (521-525). Out of these only one paper (525) has investigated the effect of simultaneous application of a small external stress during temperature cycling.

At this stage it is noted that in the nuclear industry the change in length accompanying temperature cycling of uranium, say of the α form, is referred to as "growth" while the volumetric change produced in the process is called "swelling". (The same terminologies are used even when the dimensional changes are brought about by neutron irradiation - Section 5.1.1(c).)

Presence of preferred orientation in the starting material appears to be a necessary condition for the occurrence of growth. Growth rate increased with the degree of preferred orientation but was also affected by the type of texture. Annealing of α-uranium above its recrystallization temperature or subjecting it to a β treatment made the structure more random and caused lower growth rates. A material having greater anisotropy of thermal expansion properties gave rise to faster growth rates compared with a material of lower anisotropy but of equal grain size. Growth rate per cycle increased with increasing number of cycles because with progressive deformation preferred orientation increased. Greater growth resulted when the amplitude of the temperature cycle, ΔT, was increased. Growth also depended on the holding time at the upper cycling temperature and for achieving a fixed growth rate the required holding time decreased with an increase in the upper limit of the temperature cycle. The conclusion that growth rate increases with decreasing grain size, however, is considered to be equivocal because the texture invariably changes when the grain size is altered.

When uranium rolled in the temperature range 573-773K was temperature cycled in a range above its recrystallization temperature growth rate was seen to increase slightly with an increase in the rolling temperature. The latter variable, however, had no effect when temperature cycling was done below the recrystallization temperature. Two-step rolling at successively decreasing temperatures or an increase in the percent reduction when rolling was completed in a single step also increased the growth rate. Nevertheless, the growth rate decreased sharply when the rolling temperature was greater than 873K. This was attributed to an increase in grain size and/or a decrease in preferred orientation.

In β treated uranium depending on the rates of heating and cooling either growth or shrinkage was observed. Growth rate per cycle was maximum when heating was slow and cooling was fast. Minimum growth resulted when the cycling technique was reversed. Equal rates of heating and cooling gave intermediate values for growth. When both the rates of heating and cooling were very slow, however, growth became negligible. The density of "adjusted" uranium progressively decreased with increasing number of temperature cycles. The density of high purity uranium, however, was virtually unaltered.

When a small load was applied during temperature cycling elongations in excess of 500% could be produced in "adjusted" uranium. The corresponding figure for high purity zinc was 158%. As in case of transformational plasticity the above values can be increased by carrying out the tests at constant stress instead of at constant load.

As with transformational plasticity in this case also direct linear relationships existed between the strain per cycle and the applied stress and the average strain rate and the applied stress respectively. Then, following the method of Oelschlaegel and Weiss (29) (Section 2.8) m was calculated to be unity.

When an external stress was present, the logarithm of fractional density decreased linearly with the logarithm of true strain as well as time of cycling.

<u>(c) Neutron irradiation</u>

In-pile experiments that study the deformation of metals under neutron irradiation are difficult and expensive. Therefore only a few results are available (526-528). Moreover, the rate of deformation produced by irradiation is very small and so large elongations have not been reported. In the absence of an applied stress in uranium the growth rate is zero above about 723K and maximum at about 473K. At lower temperatures the growth rate once again decreases. At a given temperature growth increases with the dose of irradiation. Change of shape in single crystals of uranium is highly anisotropic, e.g. as a result of irradiation there is an elongation in the [010] direction but in the [100] direction there is a contraction. Growth of grains in polycrystals is twice as fast as in single crystals. Moreover, the growth rate of cold worked single crystals was greater than that of undistorted specimens.

High external system pressures reduce swelling. The highly anisotropic crystal structure of α-uranium can be randomised and refined by heat treatment and alloying additions. Then the overall growth/swelling will be zero even though individual grains will continue to swell. Because of the high internal stresses, however, the material has no elastic limit and flows like a fluid even when a small external stress is applied (526). Edington et al. (129) have estimated that for the results of Roberts and Cottrell (527) m = 0.8.

5.1.2 Metallographic features

Experimental results concerning the microstructural aspects of environmentally superplastic flow are sparse. The few that are available have been reviewed below. It will presently become obvious that based on the available results no unequivocal conclusions can be drawn regarding the mechanisms responsible for environmental superplasticity.

Stobo (502) has noted that when uranium was thermally cycled through a phase transformation fine grained material coarsened but coarse grained material got refined. Based on their experimental results concerning uranium, iron, titanium and zirconium Greenwood and Johnson (150) have, on the other hand, concluded that grain growth accompanies transformational plasticity. The effect of grain size was not clear, although it was not considered to be large.

In some cases temperature cycling through a phase transformation produced no evidence for cavitation (150,502) but in others depending on the geometry of distribution of thermal stresses cavities could be observed (501).

When transformational plasticity was produced by the application of a small external stress during martensitic transformation, the martensite formed was finer. In general twins were absent. In case of the Fe-15.4% Ni alloy, regardless of whether an external stress was present or not, the martensite formed was degenerate. In the Fe-32.9% Ni alloy, on the other hand, reverse twins were always present (148). However, due to unknown reasons the application of an external stress reduced the extent of orientation multiplicity which is an important feature of martensitic transformation (509).

Edington et al. (129) have discussed the microstructural changes that accompany an increase in deformation when Bi_2O_3 and a $Zr-H_2$ alloy are thermally cycled. As the connection between these phenomena and environmental superplasticity is rather tenuous it is felt that it is not necessary to describe these results here in detail.

Results concerning temperature cycling of a material with anisotropic coefficients of expansion deal with either uranium (both high purity and "adjusted" grades) or high purity zinc. In both grades of uranium temperature cycling led to grain refinement. In "adjusted" uranium a sub-structure was also seen (524,525). On the other hand, in high purity zinc temperature cycling led to a slight increase in grain size. Limited evidence for sub-grain formation has also been reported (525). In the absence of an applied stress in high purity uranium no cavitation was seen after temperature cycling. In "adjusted" uranium, however, cavitation was present whose concentration increased with the number of thermal cycles. Perhaps there was also an inverse relationship between the concentration of voids and the grain size (521,524).

The cavity size was uneven and often comparable to that of the grain diameter. In case of "adjusted" uranium both the overall (negative) rate of density change and the time exponent of void growth increased in presence of an applied stress. In high purity zinc, however, only the (negative) rate of change of density was accelerated by the presence of an applied stress.

Optical microscopy of cathodically etched surfaces of thermally cycled "adjusted" uranium (Fig.5.6(a) and (b)) revealed that

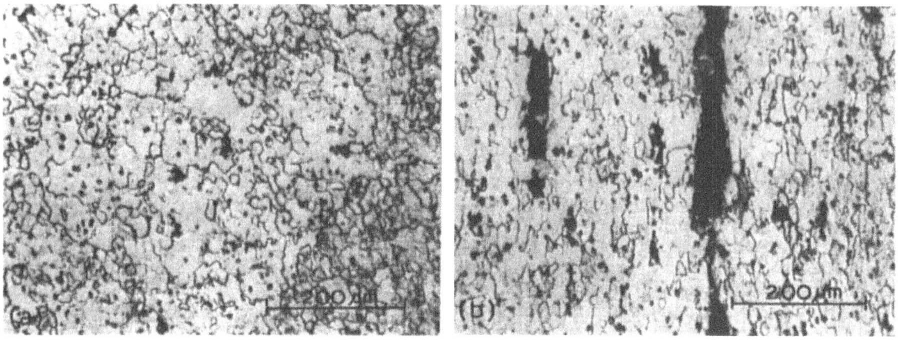

Figure 5.6 Cathodically etched surfaces of adjusted uranium showing (a) the longitudinal section of the head of a specimen after 600 cycles between 673 and 873K, and (b) elongated intergranular voids in gauge length of specimen in (a) after 385% extension. Stress axis vertical (Lobb et al. (525))

(i) the head of the specimen was substantially undeformed and almost free of porosity; (ii) the gauge length contained a large number of intergranular voids and (iii) following large elongation ($\sim$385%) the intergranular voids were considerably elongated in the direction of the applied stress. Most of the voids were found at triple point junctions, grain boundary ledges or the carbide precipitates found along the grain boundaries.

When single crystals of uranium were subjected to neutron irradiation swelling of invididual grains was caused by the formation of dislocation loops. Vacancy loops formed on planes close to (100) with a Burgers vector of a [100] and gave rise to a contraction in the [100] direction. Interstitial loops formed on the (010) planes with a Burgers vector of $\frac{1}{2}\sqrt{a^2+b^2}$ [010] and produced an elongation. The decrease in growth rate with increasing temperature (Section 5.1.1(c)), on the other hand, was attributed to the increased efficiency of clustering of point defects necessary for forming the loops (529,530).

In case of uranium irradiation did not produce any cavitation if the temperature was outside the range of 623-773K. The sub-structure produced by irradiation varied significantly with temperature and was also related to the extent of cavitation present. The cavity shape was irregular and its size ranged from 1-200 μm. When swelling increased, the cavities became larger and more numerous. Both the rates of nucleation and growth of cavities were modified by alloying additions. A critical radius was necessary for the growth of the voids and thus after an incubation period swelling became very rapid. Growth of cavities was time-dependent. Cavities were mostly present at grain and/or twin boundaries (528,531).

5.2 Theoretical Studies

5.2.1 Phenomenological approaches

Yoder and Weiss (151) have presented an analysis which is merely a re-statement of the stability criterion discussed in Section 2.5.1 for a material which exhibits both strain and strain rate hardening (cf. equations 2.15 to 2.22). They have derived the condition for stable flow as

$$\left(\frac{N}{\epsilon_t} - 1\right) d\,\epsilon_t + \left(\frac{m}{\dot{\epsilon}_t}\right) d\,\dot{\epsilon}_t > 0 \qquad .. \quad 5.1$$

which follows directly from equations 2.7, 2.10, 2.17 and 2.19. As in a tensile test $d\,\dot{\epsilon}_t$ is negative while $d\,\epsilon_t$ is positive, it has been suggested that large elongations result from the high strain hardening capacity of the material. However, as inhomogeneities in cross-section are bound to exist even in the starting material on account of limitations of specimen preparation techniques, substantial flow at high temperatures will depend to a large extent on the magnitude of m (Section 2.5). Moreover, as stated in Section 5.1.1.(a) the large elongations obtained by Yoder and Weiss (151) do not appear to be associated with environmental superplasticity because the only role of phase transformation in this case appears to be in regenerating periodically essentially strain-free austenite. Evidently, large elongation should not be employed as the sole test for the occurrence of environmental superplasticity. The large elongation ($\sim$600%) obtained in tension in case of mild steel by simultaneously applying a small rolling-type pressure that oscillates up and down the gauge length (326) is a case in point.

Gurov et al. (532) and Tikhonov et al. (533) have described a fluctuational model which applies to both diffusion-controlled and diffusionless transformations. The only difference in approach is that in the former case internal stresses do not make a large contribution to the behaviour of the "thermodynamic potential" connected with the nucleation of the new phase (as in this case relaxation processes can develop easily). On the other hand, as martensitic transformations occur at low temperatures the above assumption is not valid for this case. For the analysis it has further been assumed that the change in the "thermodynamic potential" is made up of three terms: (a) that due to the heat effect of the transition, (b) that arising from the creation of a new phase and (c) that caused by the development of a hydrostatic pressure inside the nucleus formed. It is then predicted that for maximum superplastic effects the temperature of testing should be about 0.77–0.93 of the absolute temperature of (martensitic) transformation.

The model is very vague and the physical picture is not at all clear. Moreover, detailed verification of the predicted temperature dependence of transformational plasticity is yet to be done.

Buckley et al. (501) have provided a semi-empirical expression for the growth strain during a phase transformation, G, in the absence of an external stress as

$$G = 2\,\gamma\,\left(\frac{A-B}{A+B}\right) \qquad \text{.. 5.2}$$

where γ is the linear dilatation (= $\frac{1}{3}$ x volumetric dilatation) and A and B are constants. Elaborating on the physical picture Stobo (502) has commented that in order to maintain coherency at the interface of (phase) transformation in presence of high internal stresses plastic deformation will take place on either side. A combination in one phase of high strength and high specific volume gives a distortion of one sign while the combination of high strength with low specific volume leads to a change of shape of the other sign.

A similar idea finds expression in the model of de Jong and Rathenau (27,149) which suggests that specimen elongation should be expected even at very small external stresses if these stresses remove the equivalence of the slip systems activated by the change in volume (due to phase transformation). In addition, for the martensitic case out of the many possible modes of transformation only those are considered to be feasible which allow the largest elongation in the direction of the applied stress. Then the development of mechanical weakness during phase transformation will essentially depend on the relative change in specific volume and the average yield stress of the parent and

and the product phases.

As the stress level will be limited substantially by flow in the weaker phase and the stronger phase is expected only to deform elastically, Greenwood and Johnson (150) consider that it is inappropriate to associate the strain developed during the movement of the boundary with some combination of the flow stresses of both phases. Instead they have presented a phenomenological analysis in which plastic flow is confined to the weaker phase. The form of plastic flow is considered in terms of the two extreme cases of yield and creep behaviour.

In this model (150) transformational plasticity is considered to be a pseudo-creep process (526) in which small stresses, insufficient to sustain plastic flow on their own, are able to deform a material in which an internal stress system exists. If the applied stress acts in the Z direction (concurrently with phase transformation) the increment of plastic strain, ϵ , which the weaker phase undergoes can be described by the matrix

$$\begin{bmatrix} -1/2\epsilon & 0 & 0 \\ 0 & -1/2\epsilon & 0 \\ 0 & 0 & \epsilon \end{bmatrix}$$

When the temperature of transformation is low in comparison with the melting point, time-independent behaviour is expected and the problem can be treated as one of yielding using the Levy-Mises equations for generalised stress systems. Ideal rigid-plastic response on part of the material has been assumed. If σ_o is the yield stress of the weaker phase, ΔV is the change in specific volume on transformation and V is the specific volume of the parent phase, then for the case of complete randomness (on passing both ways through the transformation temperature) the total permanent strain per complete cycle will be given by

$$\epsilon \simeq 5 \ \sigma_Z \ (\Delta V/V)/3\,\sigma_o \qquad .. \quad 5.3$$

where σ_Z is the externally applied stress.

When the phase change occurs at temperatures that are high in comparison with the melting point time-dependent flow, i.e. creep, is expected. Then starting from the commonly used modified Levy-Mises equation for creep it could be shown that

$$\epsilon \simeq 2 \; \sigma_Z (\epsilon_{I_o}) \; [5 \; N/(4 \; N+1)]/I_o \qquad .. \quad 5.4$$

where I_o is the internal stress produced by transformation, N is a constant, and ϵ_{I_o} is the creep strain that would be produced by an external stress of magnitude I_o when the internal stresses are absent.

For the case of temperature cycling of a material with anisotropic coefficients of expansion the appropriate expression is

$$\epsilon \simeq 5 \; \sigma_Z \; [\Delta \alpha \; \Delta T + (\Delta V/V)]/3 \; \sigma_o \qquad .. \quad 5.5$$

where ΔT is the amplitude of the temperature cycle and $\Delta \alpha$ is the difference in thermal expansion coefficients in the [100] and [010] directions at the middle temperature, T, of the temperature cycle. (The case of temperature cycling of α-U is the basis of this derivation.)

On the other hand, when the internal stresses are allowed to relax an additional strain of magnitude $2\sigma_Z\,[5N/(4N+1)]/E$, with E the modulus of elasticity, will result. Evidently, this component of strain is negligible in comparison with the transformational strain (equation 5.3).

The case where the transformational strain-applied stress curve extrapolates through the origin is considered by Greenwood and Johnson (150) as an example of random behaviour. The decrease in the magnitude of the strain per cycle with the addition of carbon to iron, on the other hand, has been attributed to the decrease in the maximum temperature of stability of ferrite (the weaker phase) as well as the increase in its strength.

Moreover, this macroscopic model, which applies to both diffusion-controlled and martensitic transformations, is considered to be the only realistic type of analysis possible at this stage due to paucity of critical results concerning the microstructural features of transformational plasticity (150).

Experimental results concerning phase transformational plasticity in U,Zr, Ti, Fe and Co have been summarised (see Table 4 of ref. (150)) from which it is clear that there is good agreement between the predictions of this theory and the experimental findings. Likewise, the model has been found to be satisfactory in case of a number of steels, Fe-Ni alloys and commercial grade titanium (148,498,506,511). The diverse nature of these results appears to confirm the suggestion (150) that intrinsic properties of a specific combination of crystal structures has little effect in determining the mechanism of deformation. The type of crystal structure involved simply indicates the phase in which plastic deformation is most likely (534).

For the case of $\alpha \rightleftharpoons \gamma$ transformation in steels the model predicts a larger transformational strain during heating than on cooling because in the former case the material primarily consists of (weaker) alpha. This prediction is consistent with some experimental results (29,494) but not with others (496). Moreover, Kot and Weiss (148) have shown that in case of the martensite→austenite transformation in Fe-32.9% Ni alloy the prediction differs from the experimental results by an order of magnitude. Sattler and Wassermann (509), on the other hand, have criticised this model for its failure to take into account the directional dependence of transformational strain as well as the reduction in the orientation multiplicity when martensitic transformation takes place under load. Finally, the model is silent regarding the physical processes responsible for transformational plasticity.

Chiswik (521) and Burke and Turkalo (535) have proposed a qualitative model for the change of shape accompanying the temperature cycling of a polycrystalline material with anisotropic coefficients of expansion. The driving force for the change of shape is provided by the internal stress that develops between differently oriented anisotropic grains as a result of the change in temperature. At high temperatures this stress is relaxed by grain boundary flow and at low temperatures by crystallographic deformation in some of the grains. The two mechanisms of relaxation combine to produce a "thermal ratchet" which causes continued irreversible elongation.

Contrary to experimental results this model predicts grain elongation and cavitation (in the absence of an applied stress) even in materials of high purity. Moreover, in polycrystalline materials the differential dimensional changes of adjacent grains should be prevented by the surrounding matrix.

As stated in Section 5.1.1(c) Cottrell (526) and Roberts and Cottrell (527) have analysed the flow of a fine grained random polycrystalline material under a small tensile load when it is concurrently subjected to neutron irradiation. In the absence of an applied stress there is no change in the external shape although individual grains experience growth. So long as the time of irradiation is less or equal to ($\sigma_o/E\dot{\epsilon}_g$), where $\dot{\epsilon}_g$ is the growth rate of an individual grain, accommodation will be purely elastic. For larger times plastic deformation must take place. As in this condition the material possesses no elastic limit, "liquid-like" flow is expected even when a small stress, σ , is applied. Then the rate of flow will be equal to $(\frac{\sigma}{\sigma_o})\,\dot{\epsilon}_g$.

5.2.2 Mechanistic models

Most of the theories of environmental superplasticity are qualitative. Often they are based on assumptions which are yet to receive experimental support. The theories have essentially used three ideas, viz. (a) loss of coherence between the parent and the product phases, (b) increase in diffusivity resulting from phase transformation and (c) dislocation motion.

Lozinsky and Simeonova (28), Lozinsky (536) and Arbuzova and Khandros (505) have suggested that as phase transformation involves a reconstruction of the lattice there is a partial loss of coherent bonding between the atoms. Therefore the mechanical strength of the material decreases drastically. The accompanying change in volume (on transformation), on the other hand, produces considerable internal stresses. The directional movement of atoms along the interfaces between the parent and the product phases gives rise to superplastic flow under the action of a small external stress. The above model is considered to be valid for both diffusion-controlled and martensitic transformations. According to Edington et al. (129) the low stresses, the narrow thickness of grain/interphase boundaries and the isolated nature of the transforming nuclei preclude large contributions resulting from this mechanism of flow. Sherby and Goldberg (537) using the earlier results of Sherby and Lytton (538) have, on the other hand, suggested that the observation of Lozinsky and Simeonova (28) that plasticity increased away from the centre of the specimen (which was at the highest temperature) could be due to the presence of a greater proportion of the more easily deformable alpha form of iron in these regions. There is also some evidence that during phase transformation the diffusivity of atoms is increased (539).

Clinard and Sherby (494) have proposed a mechanism which applies to the $\alpha \rightleftharpoons \gamma$ transformation in iron. As the specific volume decreases by about 1% during the $\alpha \rightarrow \gamma$ transformation, it has been postulated that vacancies are generated during the heating part of the cycle while the production of interstitial atoms is involved during cooling. As dislocation climb is usually the rate-controlling step at elevated temperatures, creation of point defects is expected to result in increased creep rates. The internal stresses accompanying the change in volume could also contribute to transformational plasticity. The direct evidence for the creation of excess vacancies during transformation (539) and the indirect evidence (540-542) that vacancy clustering occurs to relieve volume misfit produced by the formation of small precipitates (in aluminium alloys and austenitic stainless steel) appear to indicate that the above ideas could be relevant. Moreover, due to the higher energy of formation of interstitials the number of interstitials produced during cooling will be less compared with the number of vacancies produced during heating. Then the transformational strain during heating will be more than during cooling and this prediction has some experimental support (Sections 5.1.1(a) and 5.2.1). If the lifetimes of point defects are very small, all vacancies/interstitials will reach their sinks regardless of the heating or the cooling rate. Then transformational plasticity will be independent of the rates of heating and cooling. At very fast rates of heating and cooling, on the other hand, glide is considered to be the rate controlling step and on account of the decrease in the time available for glide the strain per cycle decreases with an increase in the rate of heating and cooling.

It should be noted at this stage that the excess vacancy concentration required to explain the experimental results is much less than what would be available if the entire ($\sim$1%) volume change on transformation goes to produce vacancies. Then the existence of traps to mop up a large fraction of the excess vacancies will have to be postulated in order to reconcile the experimental results with the present model. Moreover, quantitative development of these ideas is necessary before they can be taken seriously.

According to Porter and Rosenthal (510) once a relatively low threshold stress is crossed dislocations move out of their sources (under the action of a small external stress) and pile up into an equilibrium distribution at the grain boundaries. The stress concentration due to the pileups accelerate the nucleation of the new phase during transformation. Subsequently, the nuclei lose coherency with the parent lattice and are freed from their barriers. The growing interface acts as a sink as well as a source for the dislocations. Then dislocation movement is possible and rapid deformation occurs with transformation. It has also been postulated that in the early stages of transformation the effect of the applied stress is more because in this region more numerous and larger pileups which lead to greater stress concentration are encountered. The deformation will cease once the product phase becomes continuous. Porter and Rosenthal (510) themselves are of the opinion that the applicability of the model in case of martensitic transformations is less certain.

Bush and Bokros (504) have shown that the transformational strain in an oriented single crystal is somewhat greater than in the corresponding polycrystalline material. Therefore, the piling up of dislocations at grain boundaries is not an essential feature that causes transformational plasticity. Moreover, further experiments with single crystals favourably oriented for easy glide invalidated the possibility that dislocation tangles generated by prestraining could serve as the postulated barriers. Thus it would appear that transformational plasticity associated with martensitic transformation cannot be dependent on dislocations arrested at barriers of any kind. Moreover, in case of many materials the existence of a threshold stress has not been established (Section 5.1.1(a)). However, Edington et al. (129) have speculated that in these cases the threshold stress could have been supplied by the internal stresses resulting from transformation. In any case, it appears unlikely that this model will apply to those transformations in which the weaker phase is ferrite because in general there is no experimental evidence for significant dislocation activity in this component of pearlite (129).

Therefore, Bush and Bokros (504) believe that the generation of dislocations by martensitic transformation is the most important step. Subsequently, these dislocations move under the influence of the applied stress to produce a portion of the observed strain. It was also conceived that another contribution could result from the preferred operation of certain variants of the habit plane which are oriented so that the transformation shear relieves the applied stress. No metallographic evidence, however, was available to prove this latter postulate. Greenwood and Johnson (150) have noted that phase boundary slip cannot be the right mechanism because (during transformational plasticity) deformation ceases once the phase boundary stops moving. Moreover, Sattler and Wassermann (509) have observed that the changes in orientation are in the same sense during both the deformation of austenite and the deformation due to transformation. Then the dislocations will have to move on slip planes already hardened by pre-transformational flow. This is conceivable only under increasing load but hardening effects during the occurrence of transformational plasticity have been seldom reported.

Instead Sattler and Wassermann (509) suggest that transformational plasticity is the result of softening of austenite during martensitic transformation. Additional deformation will result when an external stress is applied. The major part of the transformational strain is considered to arise from continued slip on austenitic slip planes and this is made possible by a suction action of the phase boundary on the moving dislocations. This physical picture is rather elaborate and the ideas are yet to be verified experimentally. Moreover, Edington et al. (129) regard even the limited data produced by these authors as of dubious significance because the stresses used in this case pertain to the non-linear region of Fig.5.4.

Kot and Weiss (148) have proposed a dislocation model to account for the excess deformation produced by the action of a small external stress during isothermal martensitic transformation. As long as the applied stress, τ_E, is low its influence should be limited to the region surrounding the transformation front where the internal stress, τ_T, is high. By assuming that (a) the density of mobile dislocations is constant and (b) Gilman's (543) relationship for correlating the dislocation velocity with the effective stress is valid for the present case, it has been shown that when $\tau_E \ll \tau_T$

$$\dot{\gamma} = \dot{\gamma}_o \, [1 + D \, \tau_E / \tau_T^2] \qquad .. \; 5.6$$

Here $\dot{\gamma}$ is the total strain rate, $\dot{\gamma}_o$ is the strain rate due only to the presence of the internal stress and D is a characteristic drag stress. Further calculations revealed that for a given applied stress the strain rate produced was proportional to the rate of transformation. Moreover, the applied stress and the measured strain rate were linearly related so long as $\tau_E << \tau_T$ (equation 5.6). The strain (γ_E) produced by the applied stress alone was given by

$$\gamma_E = C(\Delta V/V) F \, D \, \tau_E / \tau_T^2 \qquad .. \; 5.7$$

where F is the amount of material transformed and C is a constant. Experimental verification of the model using the results pertaining to Fe-15.4% Ni and Fe-32.9% Ni alloys was satisfactory (129).

This is the only quantitative microstructural model available at present. Although Edington et al. (129) regard this approach as satisfactory for the case of isothermal martensitic transformation, it is safe to note that the two assumptions listed above are yet to receive experimental support.

5.2.3 Cavitation

The phenomenon of cavitation in environmentally superplastic alloys has not received sufficient attention. Therefore, detailed understanding is not possible. However, it is felt (129) that ideas concerning cavitation in structurally superplastic materials (Section 3.3) could be useful in this case also.

The contribution to total elongation from the formation of cavities has not been understood. Cavities produced during transformational plasticity are invariably found at the grain boundaries and in many cases they are associated with second phase particles found at the grain boundaries. Thus grain boundary sliding could be important for cavitation (Section 3.3).

Cavitation during the temperature cycling of an anisotropic material is perhaps governed by the same factors that promote cavitation in creep (521). Here again most of the voids lie along grain boundaries. The rate of growth of cavities is considered to be consistent with vacancy condensation (under the action of the internal stresses) on a fixed number of sites. In case of uranium the external stress produces additional plastic strain which gives rise to continuous nucleation of cavities. In case of high purity zinc, on the other hand, the applied stress merely increases at the beginning of deformation the number of sites available for cavity growth.

When a material is subjected to neutron irradiation, on the other hand, homogeneous nucleation of fission product bubbles takes place which coarsen under stress. Above a critical stress, however, the bubbles behave like normal cavities (531).

In summary, it could be said that both the experimental results and theoretical understanding concerning environmentally superplastic flow are rather limited, particularly when the sizeable literature pertaining to structurally superplastic deformation is viewed for comparison. In particular, experimental results on the metallographic features of deformation and cavitation are extremely sparse. Crucial experiments which could throw some light on the mechanisms of flow are yet to be performed. There is an effective continuum analysis (150) which is able to explain most of the observations. Microstructural models are mostly qualitative and are yet to receive experimental vindication.

6. Applications of Superplasticity

As described in the previous Chapters the characteristic features of the deformation of alloys which exhibit structural superplasticity is that extremely high strains can be achieved with low applied stresses with a high strain-rate sensitivity during deformation. The high strain-rate sensitivity makes these alloys resistant to localised deformation (necking or thinning). Thus it is not surprising that at an early stage it was concluded that these features could be taken advantage of in forming operations of various types. The resemblance between the behaviour of superplastic materials and thermoplastics above their glass transition temperature is clear both in their stress-strain and stress-strain rate dependence, and this suggested that forming techniques for the latter could readily be adopted for the former. Thus it was envisaged that processes such as vacuum thermoforming, drape forming, bottle-blowing etc., could be utilised for forming superplastics.

Additionally it was recognised that conventional compressive working processes, e.g. rolling and forging, could benefit from the low flow stresses associated with superplastic deformation. Then smaller working loads and lower stresses on tooling would result. There are, however, possible disadvantages. The most significant of these is the requirement to work within a limited, and relatively low, strain rate range. It has already been seen (e.g. Figs. 2.3 and 2.4) that there are three stages of superplastic deformation with the maximum superplastic response being associated with the intermediate stage. In this stage (Stage II) which normally covers a range from 10^{-4} to 10^{-2} s^{-1}, the maximum strain-rate sensitivity is observed. The actual range of strain-rates and the value of the strain-rate sensitivity index, m, in the relation,

$$\sigma_t = K\dot{\epsilon}_t^m \quad .. \quad 6.1$$

are dependent on various factors such as temperature, grain size etc. Thus it is necessary in most cases to try to maintain deformation within critical limits to optimise the superplastic response. Because of this there are economic disadvantages since the low strain rates will lead to long forming times. It is relevant to note at this stage that most common forming processes for metals and alloys operate at strain rates three to four orders of magnitude greater than those in which optimal superplasticity is observed.

Over the years the superplastic forming of a wide range of alloys has been studied. Both experimental alloys and alloys with commercial potential have been examined. The commercial alloys are expected to deal with the problem of satisfying demanding service requirements. In particular they must have resistance to creep at the service temperature. The form of the log σ_t - log $\dot{\epsilon}_t$ curves describing superplastic deformation makes it highly probable that creep properties will be poor. Two principal approaches have been followed. The first involves the use of alloying additions intended to improve creep resistance while not affecting the superplastic properties. This has had limited success (218). The second way of overcoming the creep problem is to destroy the superplastic microstructure by heat treatment between forming and service and this also has been successfully achieved (see Section 6.1.9, below).

It is also necessary to establish a combination of microstructural and deformation conditions which avoids premature rupture during forming. In particular internal cavitation must not occur.

In this Chapter we will be concerned with the application of superplasticity in forming. In doing so attention must be given not only to individual processes and theoretical analyses but also to the totality of all operations that go into producing an article under consideration. This is particularly the case when commercial developments are under examination. Then it is necessary to look at the way in which material and process costs are related. Materials exhibiting structural superplasticity will be concentrated on since although environmental (transformation) superplasticity is metallurgically interesting it does not seem to have the same potential for industrial utilisation.

6.1 Forming Operations

As described above the forming operations used with superplastic alloys fall into two broad categories, namely

(i) those processes which are derived from processes for

forming thermoplastics, and

(ii) procedures which are based on conventional working processes for metals and alloys.

A further classification is possible (544) depending upon whether or not the intention is to generate bulk changes in shape (macroforming) or to produce topographical detail (microforming) as in coining or embossing operations. In practice it is not uncommon for macroforming to precede a microforming finishing operation.

In this section we will first examine those processes based on plastics forming and then give attention to the use of superplastics in the more conventional processes.

6.1.1 Hydraulic bulging

A first demonstration of the potential of forming superplastic materials using processes normally applicable to thermoplastics was given by Backofen et al. (42) who showed that the free bulging of sheet material (the Zn-20% Al eutectoid) by pressure could be carried out such that the bulge height was greater than the blank diameter. The process is a simple one and involves clamping a circular diaphragm around its periphery and subjecting it to a one-sided hydrostatic pressure (Figs. 6.1 and 6.2). This can be achieved either by creating a vacuum on one

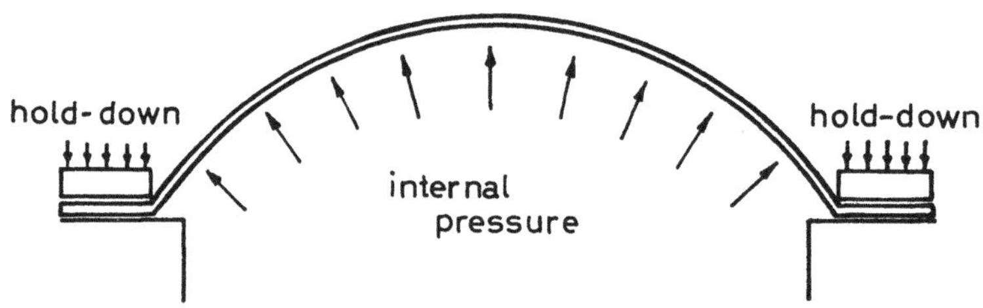

Figure 6.1 Schematic representation of hydraulic bulging

side and using atmospheric pressure to produce the deformation or by applying a fairly substantial excess pressure to one side only. This type of free pressure forming was first examined in detail by Jovane (64). He was clearly aware that ideally the deformation process needed to be carried out within a limited strain rate range appropriate to each

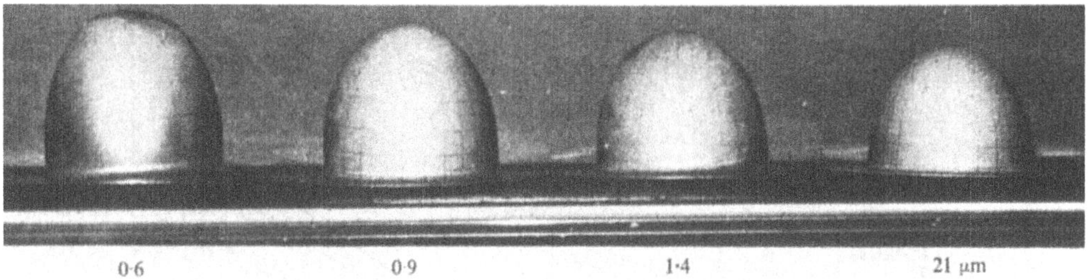

Figure 6.2 Typical forms of cups hydraulically bulged to failure. In this case each cup has a different grain size, as indicated. (Hestbech et al. (545))

material. Constant pressure forming was considered to be unacceptable due to the wide variation in strain-rate as the bulge developed while, on the other hand, constant strain-rate forming was seen to be difficult to achieve industrially because of the continuous variation of pressure which would be involved. As a result of a theoretical analysis Jovane (64) was able to outline a stepwise loading procedure which he considered to be acceptable industrially. (The various theoretical analyses for superplastic forming operations will be described in Section 6.2 after the description of the different processes is completed.) The procedure involved identifying a strain rate range within which the superplastic behaviour was good and the total forming time as short as possible. Experimental tests with tin-lead alloys showed that the bulge height increased with increasing m value and that variations in thickness after bulging increased with increasing height and decreasing m value (Fig. 6.3). These observations have since been broadly confirmed with a wide range of alloys, e.g. a zinc-0.4% aluminium alloy (189), the zinc-aluminium eutectoid (544-546), the lead-tin eutectic (65,544,547), titanium-6% aluminium -4% vanadium (548), iron-26% chromium-6.5% nickel (548) and two complex copper-based alloys (549). In particular it was found that as the m value tended to unity so thickness variations were reduced.

In many ways the most comprehensive study of bulging is that of Hestbech et al. (545) who examined in detail the effect of strain rate and grain size on performance. Performance was characterised by assessing variations in thickness and in the average meridian strain. It was found that the grain size had almost no influence on the thick-

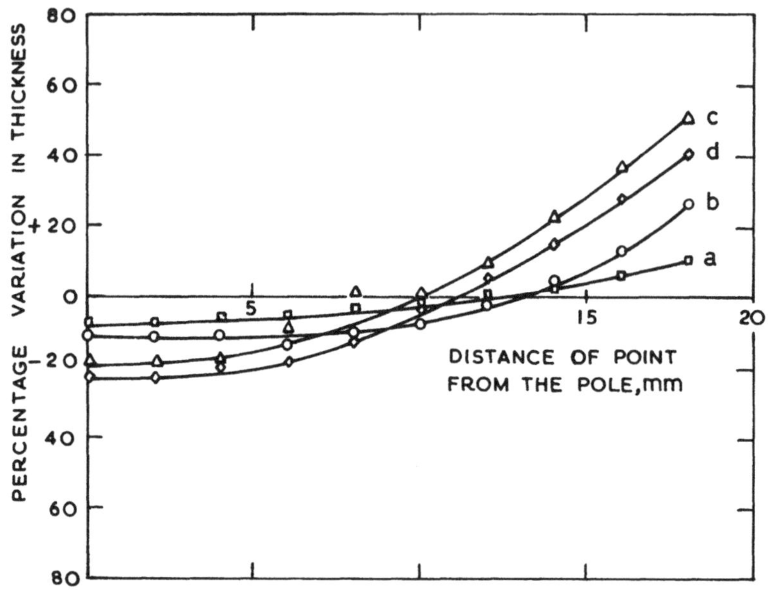

Figure 6.3 Percentage deviation of thickness from mean diaphragm thickness versus distance from the pole of bulge (Jovane (64)).
Curve a: m = 0.40, bulge height = 0.3 x diameter
Curve b: m = 0.45, bulge height = 0.3 x diameter
Curve c: m<0.3, bulge height = 0.3 x diameter
curve d: m = 0.45, bulge height = 0.5 x diameter

ness variation or on the thickness at the point of failure. It did, however, significantly affect the bulging resistance and the final surface smoothness. As far as strain rate was concerned it appeared that provided deformation was carried out in the range where m is almost constant then there is little effect. Other studies of the influence of different variables (creep-resistance-enhancing alloy additions (218), texture (189)) seem to indicate that the bulging behaviour of superplastic alloys is not complex and that in general, optimum combinations of conditions can be determined from the appropriate tensile data.

The spherical bulge test is principally a means of assessing behaviour since related industrial processes usually involve the use of a closed cavity to conform to a specified shape of the end-product. An indication of the differences in behaviour to be expected was given by Johnson et al. (550). They reported that tests in which sheets of the zinc-aluminium eutectoid (with 0.15% copper added to enhance creep resistance) were expanded into an open cylindrical cavity, showed that contact with the wall was of substantial importance. When a region of the expanding diaphragm made contact with the wall then thinning was stopped and only continued in those regions still freely expanding. This led to much less uniform thinning than that recorded with unrestrained bulging.

A further pointer to the differences to be expected in industrial practice came from work (551) with the same alloy in which bulging from a square die was examined. This work followed from earlier observations (552) that strain-hardening materials showed differences in behaviour when bulged from a square die rather than a circular one. The results indicated that, for the superplastic material under study, if the width; bulge height ratio exceeds 2:1 then excessive corner thinning occurred and the properties after deformation were poor.

In view of the above observations it is appropriate to recognise that while the free bulging of sheet can be used to assess material behaviour a more correct approach to the evaluation of sheet forming of superplastic materials is to use expansion into a closed cavity. In the main this is the procedure adopted in the plastics industry to make shapes and known as vacuum or pressure thermoforming.

6.1.2 Sheet thermoforming

After the initial bulging experiments of Backofen et al. (42), Fields (553) demonstrated the vacuum thermoforming of the zinc-aluminium eutectoid. In its simplest form, this process is one in which a sheet (preheated as appropriate) is clamped at the edges over a female mould and forced by pressure to conform to the shape of the die cavity (Fig. 6.4). It is normally used for producing low-profile parts. Fields (553) was able to show both that large draw ratios could be produced and that the material was capable of preserving detail to a very high degree.

A closely related process, known as drape-forming, makes use of a male mould with the heated sheet being drawn down over it with subsequent application of pressure to ensure that the detailed shape is developed (Fig.6.5). Both these processes involve stretching which leads to thinning but whereas with vacuum forming the thinnest regions

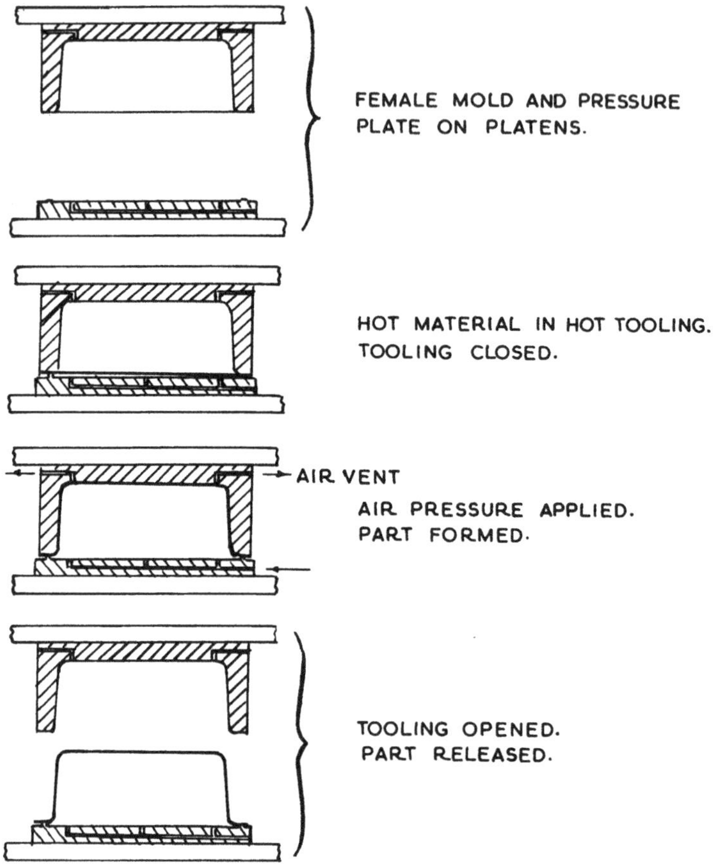

Figure 6.4 Vacuum/pressure forming of a sheet by forcing it into a female die cavity. Four part cycle begins at the top. (Davis et al. (554))

are at the centre of the piece, with drape forming the part is thinnest around the base. The need to control the distribution of thinning is a critical one and the plastics industry has developed a number of sophisticated modifications of the two basic processes in order to achieve this control, namely

(i) immediate pressure application,

(ii) billowing,

(iii) plug assist, and

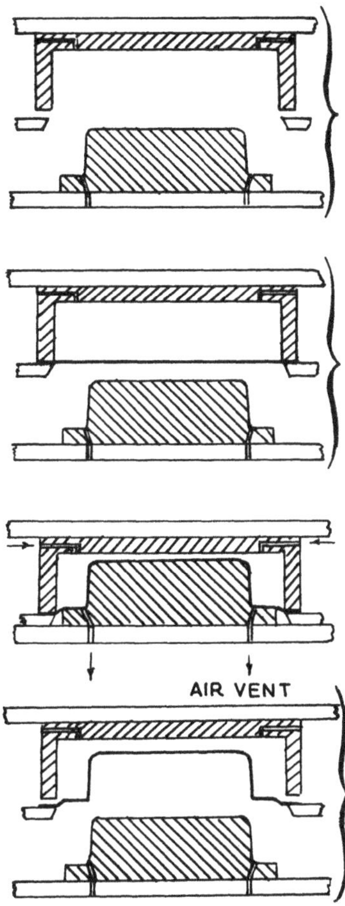

Figure 6.5 Drape forming of a sheet by forcing it over a male die. Four part cycle begins at the top. (Davis et al. (554))

(iv) limited slip drawing.

In the first of these, <u>immediate pressure application</u> is utilised to increase heat transfer and therby enhance deformation in the regions where the material is clamped. Thus, material is pre-thinned in regions where it would normally remain thick.

<u>Billowing</u> is a procedure in which sheet clamped over a die cavity and heated is first blown out of the cavity by a reverse pressure (into an empty cavity or free space) to produce relatively uniform

stretching. After the material has billowed a predetermined amount the pressure is reversed or a vacuum is applied so that the deformed sheet is then formed into the die cavity to assume the required shape.

In plug assist forming (Fig.6.6) a shaped plug is brought into

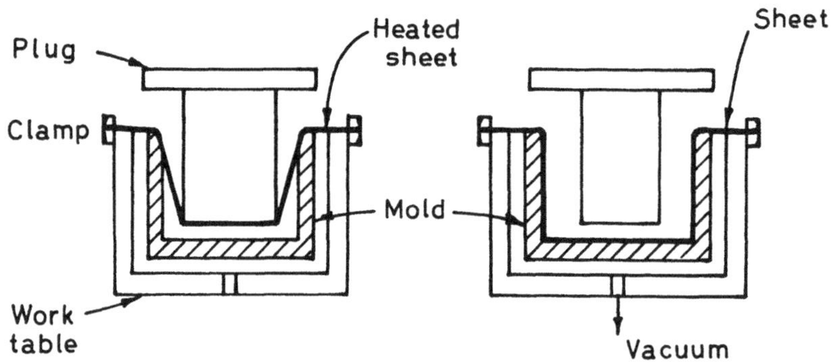

Figure 6.6 Vacuum forming with plug assist which helps overcome thinning problems in deep mould cavities. (Turner (555))

contact with the heated sheet material to produce an initial change of shape. At this stage the deformation process is similar to that occurring in drape-forming with the walls thinning more than the regions in contact with the plug. Subsequently, pressure (or vacuum) is applied in the normal way to complete forming which enhances thinning in the central regions. This stage is analogous to vacuum/pressure thermoforming. By balancing the strain in the two parts of the operation uniform thickness can be aimed for. In practice as the shapes of parts become complex it is necessary to adopt a procedure of trial-and-error to balance the strain distribution. One method of proceeding is to make use of a combination of billowing and plug assist by first billowing against a plug (or into a shaped cavity) and then completing the forming operation (Fig.6.7). With this technique the bottom of the part can be made thicker or the side walls thinner by increasing the plug penetration or by decreasing the height of the billow.

Another type of combined process is vacuum forming with snapback in which the sheet is first thermoformed conventionally by expanding into a female die cavity and then formed to final shape by being sucked back onto a male plug (Fig.6.8).

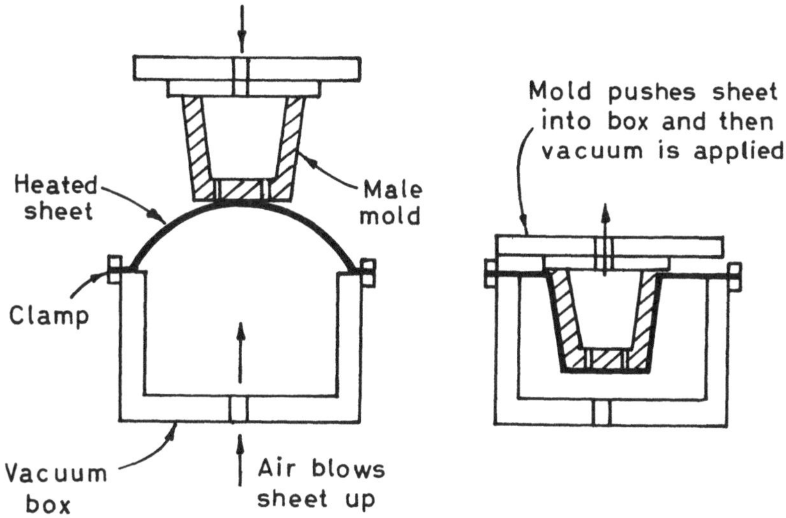

Figure 6.7 Forming with a billow plug (Mortimer (556))

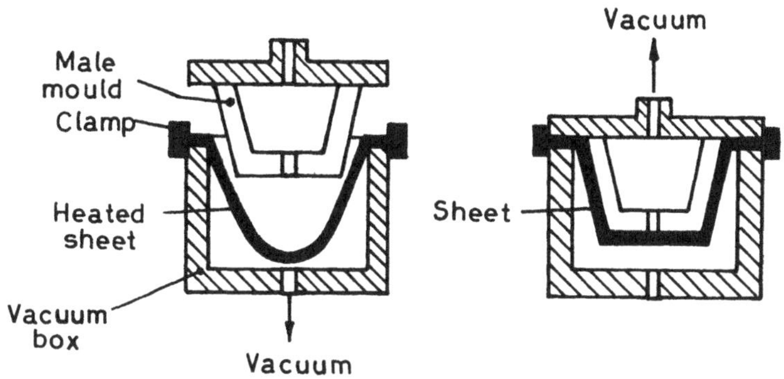

Figure 6.8 Vacuum forming with snap-back - a good method for producing parts with uniform thickness (Mortimer (556))

With <u>limited slip drawing</u> the thinning by stretching is reduced by initially clamping only lightly and allowing some drawing in of the material between the die and the clamping ring. At a later stage stretching can be re-established by increasing the clamping pressure.

As a result of the range of procedures available it does not seem necessary, as has been suggested (65), to start either with non-uniform thickness blanks or with preformed sheet in order to achieve effective forming of even quite deep parts.

The macroforming processes described above have been used successfully with a range of superplastic alloys to produce both sample commercial parts and purely experimental shapes (65,87 544,548,550,551, 557). In an early investigation Fields and Stewart (87) were able to show that additional flexibility in thickness control in simple thermo-forming could be achieved by varying metallurgical properties and pre-heat temperatures and also by using combinations of tool, die and metal temperatures to produce non-uniform temperature distributions. In the latter case this meant that different regions of the sheet exhibited different degrees of superplasticity and thinned differentially. However, it was emphasised that although the apparent versatility is desirable at first glance the consistency of dimensions on repeated cycles may be poor. On balance the evidence would seem to favour the use of the more clearly defined operations outlined above rather than resorting to localised effects as a means of thickness control.

One other advantage of the types of pressure moulding proceses already described is that because the forming operation can take place without mating male and female dies (as in conventional press forming of sheet) it is possible to design parts with re-entrant or reverse flanges. One method of tool design to achieve this is to use a flat split ring clamped over the die cavity with the inner perimeter extending inwards to the desired depth and with the required profile (Fig. 6.9). After forming is complete the split ring can be separated and removed allowing withdrawal of the part from the die. The forming of re-entrants has been studied in detail by Mamalis et al. (558), who found that there were significant variations in thickness in the side walls with lesser variations in the bulge top. Furthermore they established that with increasing forming pressures the corner radii decreased rapidly but it was essential to allow sufficient time to properly form the corners. While examining the forming of narrow grooves on the tops of the bulged parts they also observed that this required greater pressures than those needed for the more macroscopic deformation.

The ability to microform detail during the deformation of superplastic alloys was first discussed by Backofen et al. (42) who

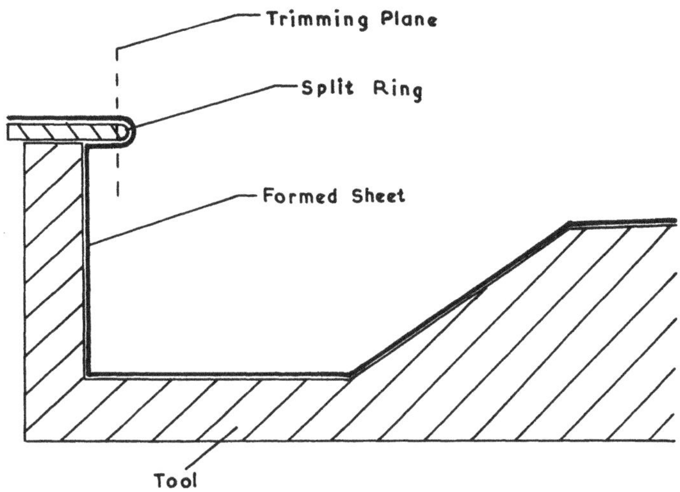

Figure 6.9 The use of a split ring to form a re-entrant part (Davis et al. (554))

bulged a sheet and forced it against a die. Thomsen et al. (544) examined microforming more fully by using a test plate (Fig.6.10) with a series of machined grooves of different geometries in its surface. They related the results therefrom to the formation of detailed embossing on more complex parts and concluded that there was no mechanical-metallurgical reason why superplastic forming should not find serious applications combining in a single design the ability to form macroscopic shape changes with the development of surface detail. This conclusion has rereceived much subsequent verification.

6.1.3 Blow moulding

This is a technique used commonly to produce plastic or glass bottles. Firstly an elongated preform is made (known as a parison) which is then expanded by internal pressure to conform to the shape of the split mould (Fig.6.11). In principle trepanned rod, seam-welded sheet or extruded tubing could also be used as preform but this requires careful attention to the clamping and sealing of seam. For bottles with complex shapes, control of wall thickness could also present difficulties. Nevertheless the feasibility of blow moulding for the formation of bottles from superplastic alloys has been clearly demonstrated

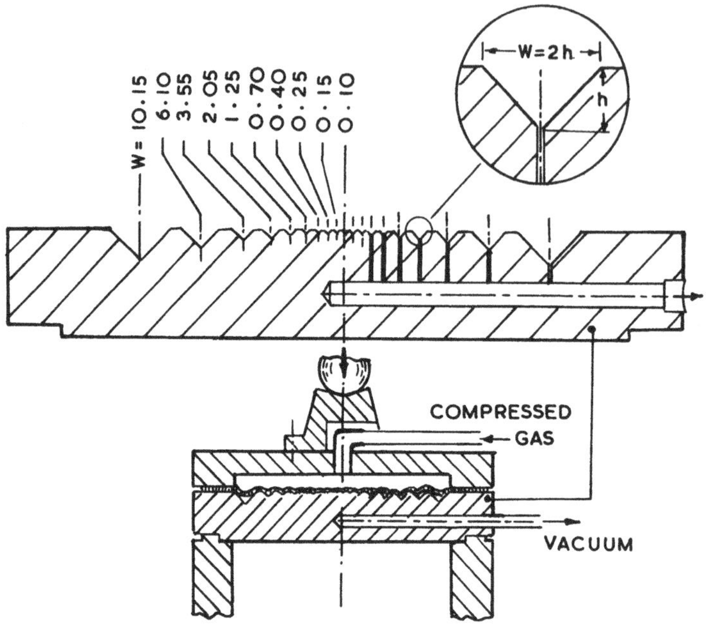

Figure 6.10 The microforming arrangement; Groove width w is in millimetres. (Thomsen et al. (544))

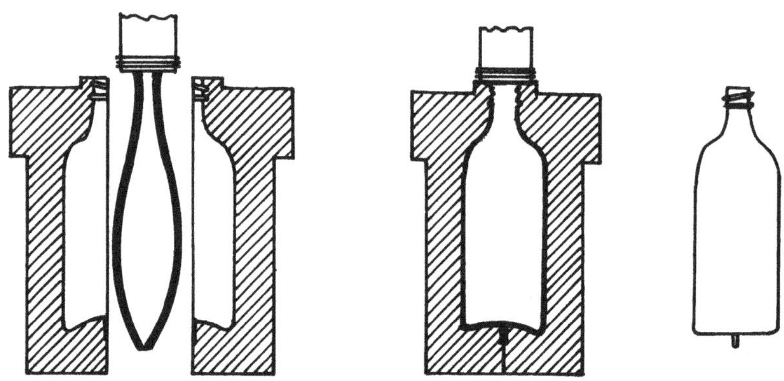

Figure 6.11 The bottle-blowing process

(65,544). Al-Naib and Duncan (65) have also shown that internal pressurization can be used to expand tubes into female dies to produce complex shapes (Fig.6.12). An ingenious variation of this latter operation

Figure 6.12 Corrugated tube formed in a single operation into a female die by internal pressurization (Al-Naib and Duncan (65))

has been described by Petersen and Lang (559) who expanded tubes of a superplastic iron-26% chromium-6.5% nickel alloy using a self-contained pressurization system. This was achieved by welding end caps on tubes which had been previously filled with powdered calcium carbonate. On heating to a temperature of ~1273K at which stage the alloy became superplastic, the expansion took place as a result of the pressure exerted by evolved carbon dioxide. In this way various configurations were formed in simple dies with excellent reproduction of surface detail.

The other forming operations that have been applied to superplastic materials are much closer to those commonly used in the metals industry for forming sheet or for bulk forming.

6.1.4 Deep drawing

In the conventional deep-drawing of sheet metals a blank lightly clamped around the edges is drawn into a die recess by a punch. The clamping pressure is normally adjusted so as to allow the blank to draw into the die but without flange wrinkling as a result of the circumferential stresses in the flange. Superplastic alloys deform at very low stresses and thus it was found in an initial series of experiments (65) that superplastic sheet of the tin-lead eutectic alloy could not be drawn successfully using standard tooling and constant pressure blank holding. At normal drawing speeds the cup wall failed prematurely before drawing was completed while at low punch speeds the cup wall elongated and thinned without drawing the flange. These failings were overcome by using a pressure-augmented deep drawing process (Fig.6.13).

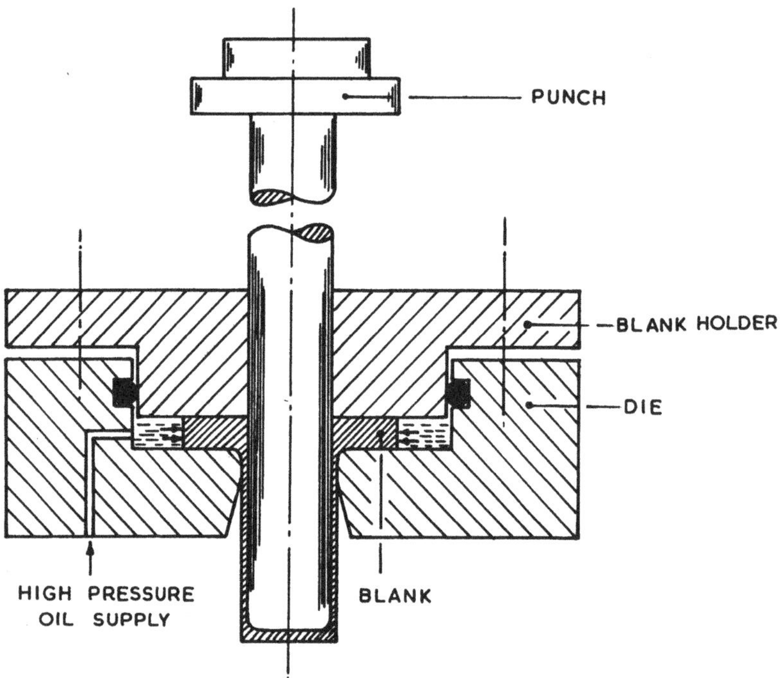

Figure 6.13 Diagram of equipment used for pressure-augmented deep drawing (Al-Naib and Duncan (65))

Cups 25.4 mm (1 in.) in diameter and 177.8 mm (7 in.) long were successfully drawn from blanks 57.2 mm (2.25 in.) diameter and 10.2 mm (0.4 in.) thick, using this equipment (65). In doing so it was recognised that the rim pressures produced a form of extrusion.

Detailed studies by Oshita and Takei (560) with zinc-aluminium eutectoid sheet and Hawkins and Belk (561) with a commercial zinc-aluminium eutectoid alloy hardened with copper and magnesium showed, however, that deep drawing was a feasible forming operation for superplastic materials provided careful process control was used. In the first study it was established that there were critical upper and lower strain rates for successful drawing. At too high a strain rate the blank failed for lack of ductility while at too low a strain rate it failed from lack of strength. In general the alloy in a superplastic condition performed better than a comparable non-superplastic alloy but even under optimal conditions the maximum drawing ratio (the ratio of blank diameter to punch diameter) was low (~2.0) and less than that obtained in the previous work (65) involving pressure-augmented drawing. The second study approached the problem much more scientifically. Hawkins and Belk (561) outlined a number of ways in which they saw that the problems of deep drawing superplastic material could be overcome including,

(a) pressure augmentation so that the metal is pushed into the die throat rather than being pulled by the punch,

(b) strengthening the metal in contact with the punch nose by simulating work hardening by using a temperature gradient (thereby maintaining the material at a temperature below the critical superplastic forming temperature so that its flow stress is higher than the material not in contact with the punch), and

(c) inducing strain-rate hardening by taking advantage of the fact that the relative velocities between the metal and the various parts of the tooling are widely variable.

Attention was concentrated on method (b) and it was shown that the drawability obtained was significantly greater than that obtained with conventional materials or with superplastic materials using other techniques. The temperatures of the punch, die and blankholder were independently controlled and the influence of the variables, forming temperature, temperature differential, blankholder load, blank diameter, clearance and draw speed and their interrelationships were established. Most notably it was reported that a maximum drawing ratio of 3.75 (in conventional deep drawing of sheet metal, draw ratios greater than ~2 are rare) was obtained at which stage the press stroke capacity was reached. The cups obtained in this work were subsequently blow moulded to form parts of intricate shape.

Clearly, modified deep drawing is an effective process for producing elongated cups and perhaps its best application will be in the manufacture of preforms for subsequent processes.

6.1.5 Punch stretching

Most sheet metal press forming operations are a combination of deep drawing and punch stretching. Behaviour in the latter process is usually assessed by forcing a hemispherical punch into a blank securely clamped around its edges. Using this type of test with varying punch speeds and different degrees of lubrication between he nose of the punch and the deforming sheet, Naziri and Pearce (189) assessed the behaviour of a zinc-0.4% aluminium alloy in the superplastic condition. Best performance was obtained at slower punch speeds with oil/polythene sheet lubrication (testing was carried out at room temperature, $0.43T_M$). Later it was shown in work with the zinc-aluminium eutectoid alloy (562) that stretching loads decreased with increasing deformation temperature and that material in the superplastic condition behaved better than non-superplastic material. This is, of course, not unexpected in view of the results reported in Section 6.1.1., when it was established that superplastic materials behaved well in hydraulic bulging, a biaxial stretching operation. The principal effect of the punch was to shift the point of failure from the pole of the dome to a point on the side wall. In these circumstances although punch stretching can be carried out with superplastic materials there would seem to be a preference for the use of free or constrained bulging.

6.1.6 Forging and stamping

Forging and stamping are closed die processes for the manufacture of precise shapes and as such are closely resembled by the compression moulding process used with plastics. Forging is usually a high temperature process and occurs under conditions in which low (but not negligible) flow stresses are encountered. Then geometrical and frictional effects play a significant role and these often combine in such a way as to inhibit flow and give rise to incomplete filling of the die. Superplastic alloys have very low flow stresses at low or intermediate strain rates and although compression tests have shown (87,99,563) that frictional constraints can lead to pressure build up, the effect is much less marked than in compressive deformation of conventional alloys. Thus it can be anticipated that if sufficient time is allowed then the use of superplastic alloys in closed die forging should result in excellent die filling with efficient reproduction of very fine detail. Compression moulded products were first shown by Fields and Stewart (87) but the first extensive study of the closed die forging of superplastic alloys was presented by Stewart (564). He performed a range of forging and upsetting tests on a copper-hardened superplastic zinc-aluminium eutectoid alloy making use of a number of forging temperatures and speeds. His work established quite clearly that excellent die filling could be obtained even in very thin web sections and that sharp corners could be produced and small draft angles used. (Prior less detailed work (563) had indicated that high precision was obtainable at low strain rates but not at the high strain rates and lower temperatures used in stamping.) Other results (105) of the same period provided additional confirmation of the high forgeability of superplastic alloys.

For the most effective use of superplastic alloys in precision forging it is necessary to reduce the forging cycle time as far as possible. One approach to this is to use a two-part cycle comprising a period of constant and rapid ram speed followed by a dwell at constant load. This procedure has been used most successfully for the precision forging of a modified superplastic zinc-aluminium eutectoid alloy (565). The cycle is shown in Fig.6.14 and some typical parts in Fig.6.15.

To some extent the forging of superplastic alloys is similar to creep forming (566,567) although that process is much slower. Both suffer from the disadvantage that the tools need to be heated thus offsetting to some extent the gains arising from the reduced forging loads.

One other way in which superplastic forging can be used effectively is in producing dies by coining or hobbing, for use in subsequent operations. Coining dies, made by forming detailed impressions in superplastic blocks at low strain rates, have been used to advantage in subsequent stamping operations at high strain rates (563). The other application envisaged (554) is the manufacture of tooling for plastics moulding by forcing a heated blank of superplastic metal over a master pattern. The process is illustrated in Fig.6.16. This has several advantages, e.g. reduced costs, ease of fabrication of male master etc., but as yet the process is not fully evaluated although examples have been described in the literature (568).

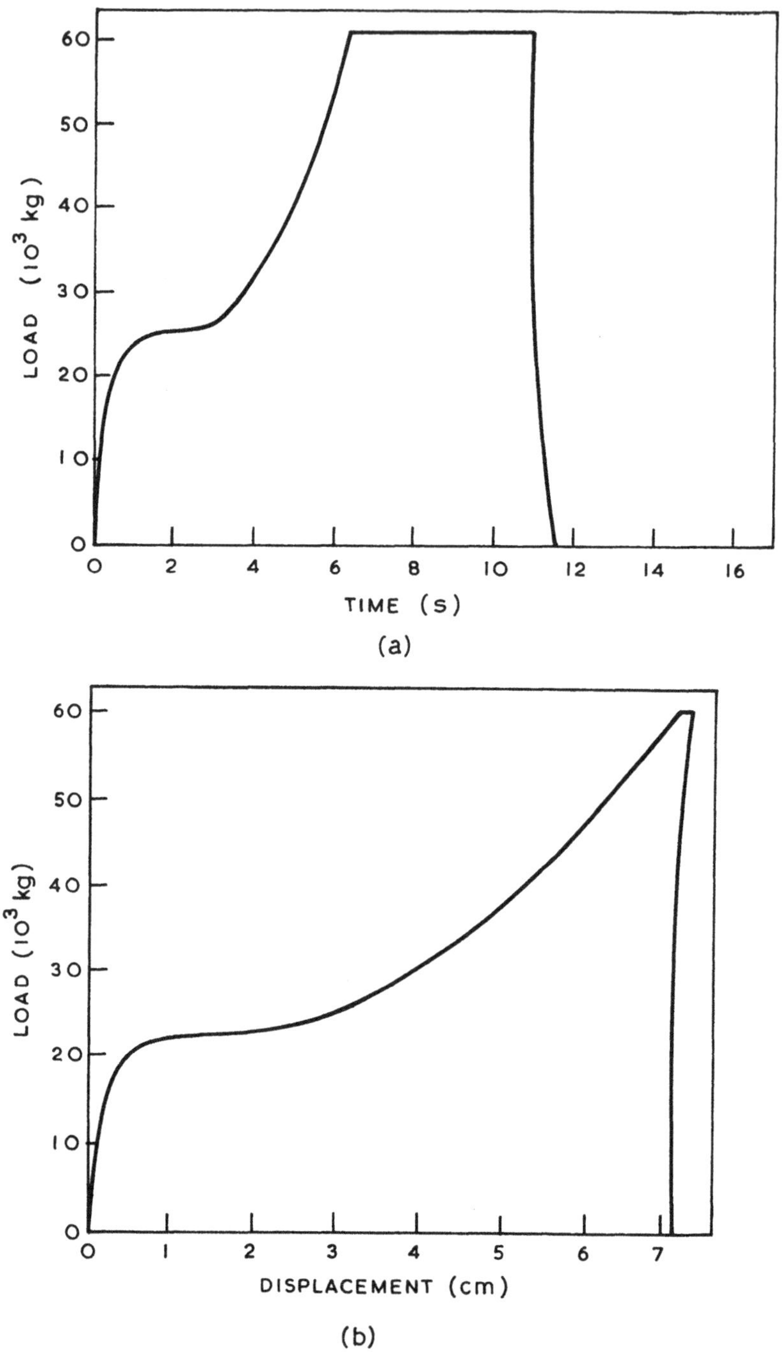

Figure 6.14 Duplex cycle for precision forging of a superplastic alloy (a) load as a function of time, (b) load as a function of displacement (Balliett et al. (565))

Figure 6.15 Precision-forged parts made from a modified zinc-aluminium eutectoid (a) cups formed from billets of various thickness, (b) castellated and hexagonal heads, (c) a part showing as-forged threads (Balliett et al. (565))

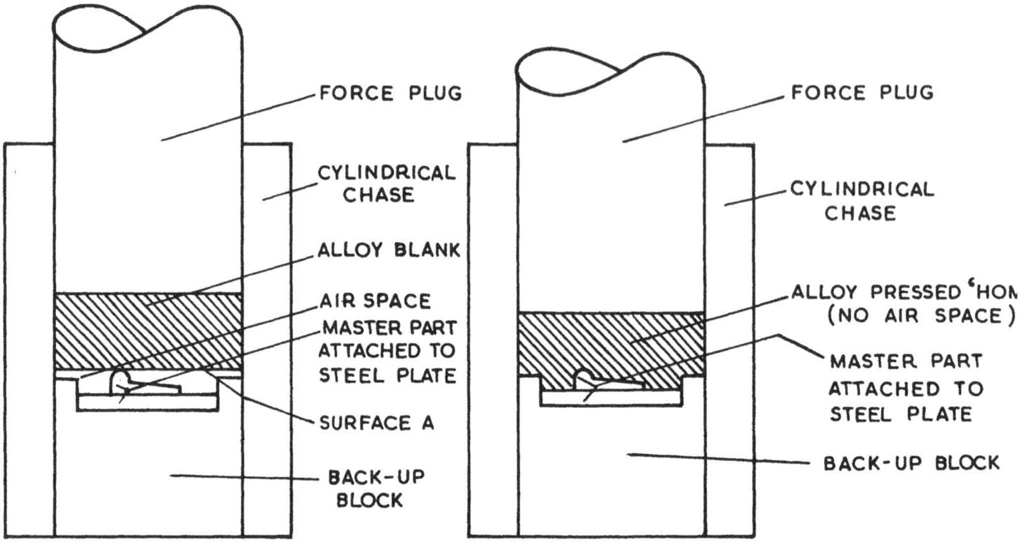

Figure 6.16 Process for the production of dies for plastic moulding by indenting a superplastic alloy blank (Davis et al. (554))

So far in this Section we have concentrated on the more detailed information relating to the forging of superplastic alloys. It is clear that a much more extensive range of alloys has been evaluated empirically (see, for instance, Hubert and Kay (568)) including titanium alloys, although details are not readily available. Here we are approaching commercial developments and these will be considered more comprehensively in Section 6.3.

6.1.7 Extrusion

A number of successful programmes involving the extrusion of superplastic alloys have been undertaken. The first reported study (569) was aimed at investigating flow behaviour for comparison with the behaviour of conventional alloys during extrusion. Particularly it was intended to determine if, and what, differences existed in their different velocity and stress fields. The work was carried out with the Sn-Pb eutectic using a technique (570,571) known as "visioplasticity". As a result it was concluded that differences were minimal for extrusion at different speeds and with a range of extrusion ratios. Such slight variations as were recorded were attributed to small differences in boundary friction. This work was followed up by a brief study by Tang (572) primarily aimed only at evaluating some theoretical predictions (see Section 6.2.2). He showed that at low stresses and extrusion velocities the superplastic material (the Sn-Pb eutectic) behaved as a pseudoplastic fluid but not at high stresses and velocities when something akin to plug flow occurs.

At almost the same time Oshita and Takei (573) in a companion study to those referred to above for deep drawing (560) and punch-stretching (562) showed that for the zinc-aluminium eutectoid a decrease in the extrusion pressure resulted from an increase in temperature although, provided $m > 0.45$, there was no significant dependence on extrusion speed. Furthermore, it was established that material in the superplastic condition was always more readily extruded than non-superplastic material under similar deformation conditions. To support this Belk (163) while principally concerned with the optimisation of production routes for superplastic zinc-aluminium alloys, produced evidence that confirmed that material in the superplastic condition is amenable to extrusion requiring only about half the load of non-superplastic alloy. Belk also confirmed a previous observation (65) that back extrusion to form tube preforms was a viable operation with superplastic materials.

The results described give a firm indication that the extrusion behaviour of superplastic alloys is good even though little effort has so far been given to taking advantage of the lower deformation stresses. Probably the most pointed observation was that of Oshita and Takei (573) who recorded that the best behaviour occurred when in the range of optimal superplasticity.

6.1.8 Dieless drawing

Because of plastic instability (necking) tensile forces are little used in metal forming. Johnson (574) devised an elegant way of taking advantage of the stability of superplastic metals in tension by proposing that a bar could be extended progressively by restricting the strain to a locally-heated zone which was made to traverse along the bar length (Fig.6.17). This process was referred to as die-less drawing since the reduction in area in the heated zone occurred in much the same way as the deformation in the die in conventional wire drawing. The desired cross-sectional area is determined by the relative values of the drawing speed (V_D) and the rate of traverse of the heating coil (V_R)

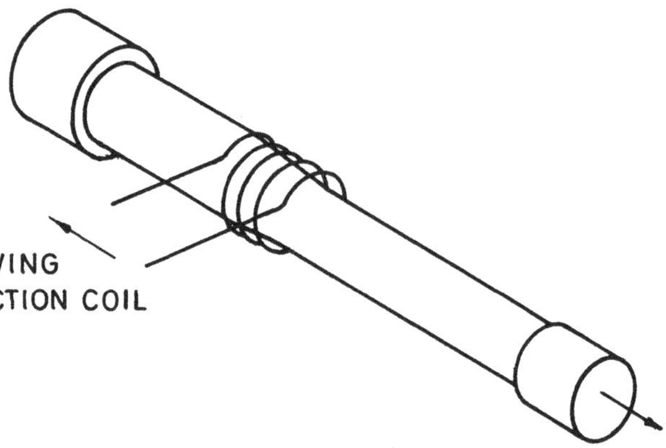

Figure 6.17 Schematic illustration of die-less drawing (Johnson (574))

and is given by

$$\% \text{ reduction in area} = \frac{100\ V_D}{V_D + V_R} \qquad .. \quad 6.2$$

Experimental work showed that reductions of up to 83% were possible using this technique. It is also possible in principle to adopt the technique to produce controlled tapering or stepped bars. As a process it is also unusual in that it is one of the few (perhaps the only) which are capable of adaption for use with alloys exhibiting environmental superplasticity.

6.1.9 Powder metallurgy processes

As described in a number of other places in this monograph an ultra-fine grain structure is a principal requirement in alloys which are to exhibit structural superplasticity. Additionally the alloys should be compositionally homogeneous on average even when they are dual- (or multi-) phased and contain a range of alloying additions. As proposed by Nicholson (291) this should make them a natural target for powder metallurgy. That this was clearly so was quickly confirmed by the reports relating to the work on the Gatorising process (see Athey and Moore (575) and Gessinger and Bomford (576) for details) which took advantage of the superplastic behaviour exhibited by compacted powder preforms of nickel-base super alloys (283). In this process an ultra-fine grain size was obtained in compacted powder by extruding near but below the recrystallisation temperature of the alloys. Adiabatic heating during extrusion initiates recrystallisation which produces the required grain structure for superplasticity. The alloy can then be forged at low strain rate to near finished size. A subsequent high temperature annealing process is used to dissolve stabilising precipitates and to promote grain growth thus eliminating the superplastic response. This is followed by an ageing treatment to develop optimal in-service properties. The sequence of events is illustrated schematically in Fig.6.18. This is unquestionably one of the most sophistica-

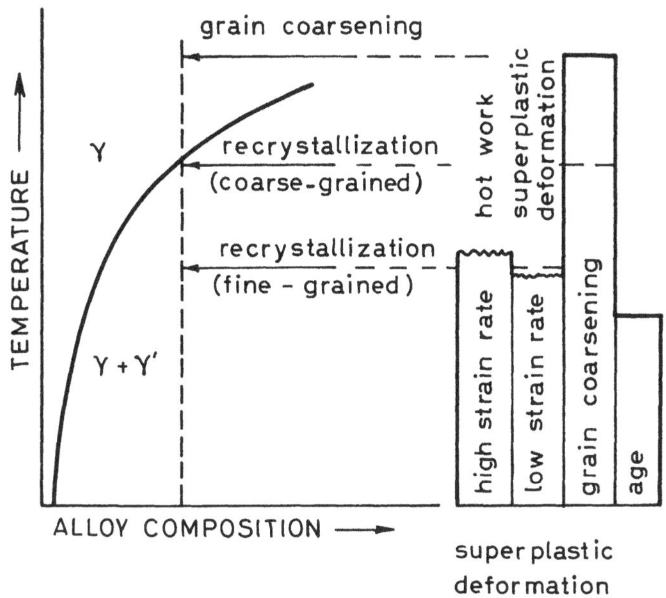

Figure 6.18 Schematic representation of the thermomechanical processing schedule for forming superplastic powder-consolidated super-alloys (Gessinger and Bomford (576))

ted and successful applications of superplasticity in forming.

6.2 Theoretical Analyses of Superplastic Forming Processes

Most theoretical analyses have characterised the material

behaviour by assuming that it can be described by equation 6.1. Only Tang (572) and Tang and Robbins (547) have considered a more complex relationship making use of the equation (316)

$$\dot{\epsilon}_t = \frac{A\sigma_t^2}{\underline{L}^3} + B\sigma_t^2 \sinh(\beta\sigma_t^{2.5}) \qquad .. \quad 6.3$$

where A, B and β are material constants at a given temperature and L is the grain size. The analyses have only dealt with a very limited range of processes with most attention being given to deformation by hydraulic bulging (see Fig.6.1).

6.2.1 Analysis of hydraulic bulging

In defining the problem Jovane (64) proposed that since the strain rates appropriate to superplastic deformation are low and vary only gradually the inertia forces can be neglected. Thus the state of stress will be determined by the equilibrium conditions, the material properties and the initial and the final boundary conditions. He made the following assumptions,

(i) the material is isotropic,

(ii) the volume remains constant,

(iii) elastic strains are negligible,

(iv) the material does not work-harden and the elastic limit is so low that it may be neglected,

(v) the thickness : diameter ratio of the bulging diaphragm is very small so that bending effects are negligible,

(vi) at any instant the membrane is equivalent to part of a thin sphere subject to internal pressure, so that the curvature and thickness are uniform and a state of uniform biaxial tension exists at any point on the membrane,

(vii) at the periphery the diaphragm is rigidly clamped and deforms as a frictionless hinge, and

(viii) the equation of state can be expressed in terms of

equivalent stress and strain-rate as a modified form of equation 6.1 namely,

$$\bar{\sigma}_t = K\bar{\dot{\epsilon}}_t^m \quad .. \ 6.4$$

where equivalent stress and strain-rate are given by the expressions

$$\bar{\sigma}_t = [\tfrac{1}{2}((\sigma_1-\sigma_2)^2 + (\sigma_2-\sigma_3)^2 + (\sigma_3-\sigma_1)^2)]^{\frac{1}{2}} \quad .. \ 6.5$$

and $$\bar{\dot{\epsilon}}_t = \left[\tfrac{2}{9}((\dot{\epsilon}_1-\dot{\epsilon}_2)^2 + (\dot{\epsilon}_2-\dot{\epsilon}_3)^2 + (\dot{\epsilon}_3-\dot{\epsilon}_1)^2)\right]^{\frac{1}{2}} \quad .. \ 6.6$$

respectively. In equations 6.5 and 6.6 the σ_i and $\dot{\epsilon}_i$ are true principal stresses and strain-rates, respectively. The geometry was taken to be as shown in Fig.6.19 where s is the current thickness of the diaphragm,

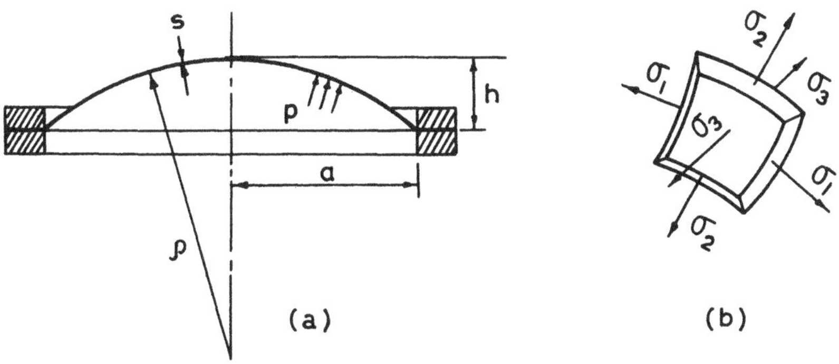

Figure 6.19 Diagram of a section of the bulged diaphragm and the stress system on an element (Jovane (64))

h is the polar deflection, a is the die radius, ρ is the instantaneous radius of curvature and p is the hydrostatic pressure. During the analysis it was found necessary to introduce a non-dimensional height H ($= \frac{h}{a}$) with the initial constraint that at time t=0

$$H = H_o \neq 0 \quad .. \ 6.7$$

For deformation at constant pressure the analysis yielded the following equations relating,

(a) current strain rate to initial strain rate, i.e.

$$\frac{\dot{\bar{\varepsilon}}}{\dot{\bar{\varepsilon}}_o} = \left[\frac{H_o}{H} \frac{(1 + H^2)^2}{(1+H_o^2)^2} \right]^{1/m} \qquad .. \; 6.8$$

(b) non-dimensional time, $\tau \left(= \frac{\dot{\bar{\varepsilon}}_o}{2} \left(\frac{H_o}{(1+H_o^2)^2} \right)^{1/m} t \right)$ and

non-dimensional height, as

$$0<H<1: \quad \tau = 2 \left[H^{(3-b)/2} \sum_o^\infty n \binom{b}{n} \frac{H^{2n}}{4n+3-b} - H_o^{(3-b)/2} \sum_o^\infty n \binom{b}{n} \frac{H_o^{2n}}{4n+3-b} \right] \qquad .. \; 6.9$$

$$H>1: \quad \tau = 2 \left[H_o^{(3b+5)/2} \sum_o^\infty n \binom{b}{n} \frac{H_o^{-2n}}{4n-3b-5} - H^{(3b+5)/2} \sum_o^\infty n \binom{b}{n} \times \frac{H^{-2n}}{4n-3b-5} \right] \qquad .. \; 6.10$$

where $b = -\left(\frac{m+2}{m}\right)$ and $\binom{b}{n} = \frac{b\ (b-1)\ \ldots\ [b-(n-1)]}{1.\ 2.\ 3.\ \ldots\ n}$.

Equations 6.9 and 6.10 are not valid for H=1 and, therefore, equation 6.9 was extrapolated to bridge the gap. Calculations based on equation 6.8 showed that the strain rate could be expected to vary by up to three orders of magnitude during bulging to a hemispherical form.

If the deformation is considered to occur at constant strain-rate then a second set of equations can be derived, viz.

(a) that relating the initial and the current pressures, i.e.

$$\frac{p}{po} = \frac{H(1+H_o^2)^2}{(1+H^2)^2 H_o} \qquad .. \; 6.11$$

and

(b) the non-dimensional time and height parameters,

$$\tau = \frac{1}{2} \left[\frac{H_o}{(1+H_o^2)^2} \right]^{(1/m)} \ln \left[\frac{1+H^2}{1+H_o^2} \right] \qquad .. \; 6.12$$

Jovane (64) realised that the technological forming of superplastic materials was unlikely to occur at constant pressure or constant strain rate. Thus it was proposed that the deformation path be split into a series of constant pressure intervals chosen so that the strain rate will be maintained within a predetermined range between the start and the end of the interval.

Tabulated results were then presented (see the original paper) relating the changes in non-dimensional height, H_i to H_f, non-dimensional time, τ, and a dimensional pressure parameter, P, given by

$$P = \frac{a}{4S_o K\bar{\dot{\epsilon}}^m} \cdot p$$

across the interval for a given value of m (in the range 0.35 to 1.00) and for given strain rate limits $\bar{\dot{\epsilon}}_i$ and $\bar{\dot{\epsilon}}_f$. The practical application of this procedure was seen as involving the use of a knowledge of the superplastic material under test to define a strain-rate range in which the superplastic behaviour was good, the forming time short and over which m was effectively constant (and as high as possible). The values of the pressure path could then be determined from the tabulated values.

The comparison of theoretical predictions with experimental measurements only involved constant pressure deformation and it was concluded that the agreement was satisfactory. This was despite the fact that other measurements showed that the thickness did not remain constant which clearly invalidates an essential part of assumption (vi).

Later, two independent but similar analyses (548,577) tackled the problem of thickness variations. Apart from allowing the thickness to vary from place to place in the diaphragm both analyses involved other assumptions similar to those of Jovane (64) including a need to assume an initial radius of curvature. Each analysis led to a set of equations which needed to be solved iteratively by computer.

Holt's treatment (577) (which was closely related to the procedure of Woo (578) who analysed the bulging of a plastic circular membrane) involved dividing the diaphragm into a series of ten equal-width annular zones and following the successive deformation of the zones working from the outer periphery inwards. With reference to Fig.6.20 the ruling equation relating the incremental change in the mean radius of an annulus to an increment of time is,

$$\frac{\partial \rho}{\partial t} = \frac{D}{4}\left[\frac{pD}{4S_oK}\right]^{1/m}\left[\frac{2\rho}{D}\right]^{(3/m)+1}\left[\frac{2\rho_0}{D}\right]^{-2/m}\sin^{-1/m}\theta \quad .. \quad 6.13$$

where K and m are the constants in equation 6.1. Knowing $\delta\rho$ the new

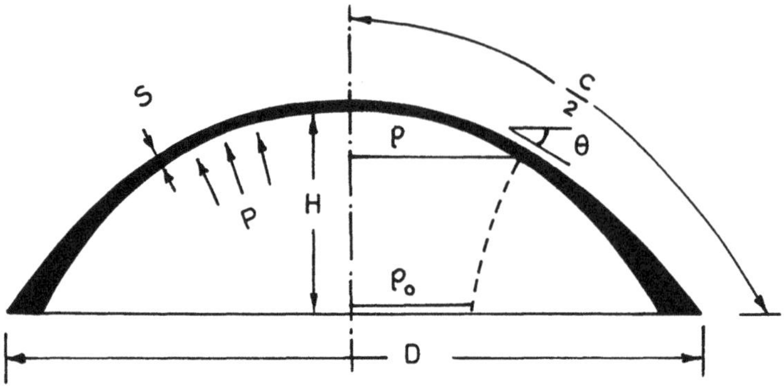

Figure 6.20 Bulge and groove geometry used for the analysis of microforming (Holt (577))

thickness and width of the annulus can be determined from the relations (based on the assumption of balanced biaxial stress and strain),

$$\frac{S}{So} = \left[\frac{\rho_o}{\rho}\right]^2 \quad .. \; 6.14$$

and

$$\rho sw = \rho_o s_o w_o \quad .. \; 6.15$$

where w and w_o are the initial and final width. Since the outer edge is fixed a knowledge of ρ and w for the outer annulus fixes that part of the profile and locates the outer edge of the next annulus. Thus by working from the outside to the centre both a bulge profile and a thickness profile can be determined for successive increments of time. Although the analysis was applied to the formation of a bulge under constant pressure it could readily be adapted to deal with stepwise pressure loading. Comparison of predictions based on this analysis with experimental data from Thomsen et al. (544) showed that the agreement was best at higher m -values.

The above analysis although comparatively straightforward suffers from the disadvantage of the assumption of balanced biaxiality which is clearly in error at the outer edge where clamping imposes plane strain conditions. This was dealt with by Cornfield and Johnson (548) by incorporating the theory of Timoshenko (579) which allows the dome to remain spherical but assumes that bending moments exist at the clamped edges. As a consequence the formulation is more complex than with the two analyses discussed above. Nevertheless the correlation of predictions with experiment was good.

Other predictions (547) do not improve the position. For practical applications what is needed is an ability to make predictions of acceptable accuracy with a minimum of computation. For a first estimate the analysis of Holt (577) would seem the most viable. Certainly in view of the probability that the manufacture of real parts from sheet will involve much more complex operations than simple bulging (e.g. as described in Section 6.1.2) a more sophisticated approach is not justified.

6.2.2 Analysis of bulging into a 90° V-groove

This microforming operation (see Section 6.1.2, especially Fig.6.10) has also been analysed by Holt (577). The geometry and nomenclature he considered is shown in Fig.6.21 and the analysis yielded the

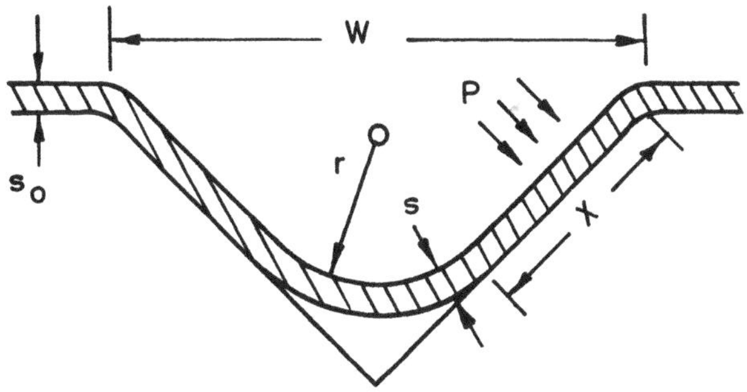

Figure 6.21 Bulge and groove geometry used for the analysis of microforming (Holt (577))

relation,

$$\frac{s_o^{1.38}}{rw^{0.38}} = \left[\frac{p}{K} \cdot t^m \cdot \left(\frac{2.65}{m}\right)^m \right]^{1.38} \quad \ldots\ 6.16$$

This was shown to be a reasonably good description of the problem by comparing equation 6.16 with results from Thomsen et al. (544). Subsequently additional confirmation was provided by North (580). Although it was postulated that the analysis could be extended to grooves of different angle this extension has not been reported.

6.2.3 Analysis of extrusion

Although this is the only other forming process for superplastic materials which has received attention, the analyses (572, 581) are based on the assumption of constitutive relations which are such as to make their validity highly questionable. Tang (572) assumed equation 6.3 and considers steady extrusion through a narrow rectangular orifice and through a circular orifice. He developed complex power series expansions capable of predicting velocity profiles and bulk flow rates. Of these the most interesting have been referred to earlier (Section 6.1.7) and relate to the prediction that only at low extrusion pressures can the material be classed as a pseudoplastic fluid. This conclusion, however, does largely follow from the nature of the assumed constitutive equation.

The other work (581) is principally aimed at comparing two different analytical procedures and does not contribute significantly to the understanding of the behaviour of superplastic materials during forming by extrusion.

6.2.4 Regression analyses

One problem facing all those involved in the theoretical analysis of superplastic forming operations is the assumption of a suitable constitutive equation to describe the material behaviour. This assumption is complicated by the recognition that the K- and m-values in equations like equation 6.1 are only approximately constant and depend on the values of grain size, strain rate etc. One approach that has been presented (582) involves the use of experimental data for the derivation of multidimensional regression equations using a generalised polynomial series. The method is a complicated one and although it gives rise to manipulable equations as well as an optimization procedure its real range of usefulness will be limited by the approximations necessarily involved in most theoretical studies (into which the prediction equations generated by the analysis will be incorporated).

6.3 Practical Developments

So far we have been concerned with fairly basic principles

relating to the application of superplasticity in forming processes. Descriptions have been given of the use of experimental alloys to assess feasibility in different operations but most of the work referred to was exploratory. It is now appropriate to concentrate attention on practical developments relating to alloys and processes where the approach has been very much commercially oriented. While examining practical developments it is necessary also to consider the design of parts for superplastic forming and tooling for production. Before embarking on these detailed examinations it is important to list the advantages and disadvantages associated with superplastic forming. This is done in Table 6.1.

In the end economic factors will determine the range and nature of the alloys and processes that achieve commercial viability. However, it will not be possible here to give detailed attention to cost analyses because of their necessarily proprietary nature.

Nevertheless it is possible to make some broad generalisations with regard to cost, placing particular emphasis on the <u>total</u> of costs involved in production. For example, it is widely recognised that in conventional press-forming of sheet for the automobile industry, although materials costs are low, the costs of tools and presses are high. This is particularly the case with automobile pressings since use of sheet steel requires that the final part be produced in a sequence of operations. Thus the production of a typical body panel might involve four or five steps such as blanking, forming, pierce and trim, restrike and flanging. Each of these steps would require its own expensive tooling. In addition, with regard to advantage 4 in Table 6.1, in press forming about 40 percent of the steel brought in is returned as scrap (583). Some early comparative cost figures indicated that for a steel pressing the costs were distributed in the approximate ratio 1 : 1.3 : 3.6 for tool amortisation, materials and labour, respectively, whereas for a superplastic alloy the corresponding proportions were 0.4 : 3.6 : 1.9. Additional savings can accrue because residual stresses are reduced and problems such as elastic springback and distortion are virtually eliminated. In the end a true evaluation can only be made by concentrating on a single pressing or a simple assembly and taking into account not only the factors referred to above but also related factors such as ease of assembly, overall weight savings etc.

TABLE 6.1

Advantages and disadvantages of superplastic forming (Edington (80))

Advantages:

1. Vastly improved formability is obtained compared with conventional forming processes even for high-strength low-ductility materials. Thus processes requiring large amounts of deformation can be performed in one operation, often producing components with complex shape. As a result, multistage manufacturing processes that involve extensive welding or machining may be bypassed, or previously unattainable shapes may be produced.

2. Forming pressures are drastically reduced; in some cases enough to make feasible the use of cheaper, novel, or more lightweight, forming equipment.

3. Considerable cost savings may be made mainly because extremely close tolerance can be guaranteed which reduces machining costs and the process is not, therefore, labour intensive.

4. Wastage is minimized so that there is maximum utilization of material, which is an important saving in energy intensive materials such as Ti or Al alloys.

5. A uniform microstructure is produced which leads to uniform, reproducible, mechanical properties throughout the body of the finished product.

6. Service properties are generally improved over conventionally formed material because a fine, uniform grain size is obtained which leads to better strength, ductility, and fatigue resistance.

Disadvantages:

1. The creep properties of the component are often inferior to those obtained by conventional forming routes, primarily because of the fine grain size.

2. The process is not, in general, suitable for rapid mass production techniques although reasonably short forming times are possible. Consequently, superplastic forming is only feasible for low-to-medium batch production.

6.3.1 Alloys

Although a very wide range of alloys has been shown to be superplastic (see, for instance, Table 3.1), the list of commercially viable alloys as evidenced by patent applications is much smaller. (See the Appendix to the paper by Edington (80) for a list of major patents up to the end of 1974.)

There are three principal classes,

(i) low temperature alloys based on the zinc-aluminium system,

(ii) intermediate temperature alloys of aluminium, and

(iii) the higher temperature alloys, principally stainless steels, titanium alloys and nickel-based superalloys.

In the present context we will examine the development of alloys of commercial importance together with their limitations, design procedures etc., by paying attention to representatives from each of these groups.

Zinc-aluminium alloys

The first commercial developments in which superplastic alloys were used to form machine parts involved the use of the zinc -22 wt.percent aluminium eutectoid. Hundy (583) and North (557) described applications of this alloy (given the trade name "Prestal") in autobody construction and showed that complex assemblies were capable of being fabricated and that the alloy could be joined by a variety of welding processes. They also showed that the alloy had good corrosion resistance (although it is somewhat susceptible to intercrystalline corrosion) provided care was taken to avoid dissimilar joints with mild steel. Subsequently Fields and Stewart (87) also illustrated the fabrication of parts with this alloy. Unfortunately in the absence of alloying additions the zinc-aluminium eutectoid alloy exhibits poor room-temperature creep resistance and as a consequence derivative alloys with superior performance were sought. The most successful results were obtained with small alloying additions (usually less than $\sim$ 1%) of one or more of the elements, magnesium, copper, titanium or manganese, frequently with off-eutectoid base compositions. Addition of copper and magnesium improved also the resistance to impact and attack by moisture (219). When the copper content was about one atomic percent intergranular corrosion was virtually eliminated (584). The presence of small amounts of alkali and/or alkaline earth metals also improved the resistance to inter-crystalline corrosion (585). This approach gave rise to a series of commercial alloys best known of which are those with the trade designations ZAM (586,587), SPZ (588) and HSZ (565). These alloys require careful production processing (163,362) and this can best be illustrated by considering the range of factors that can influence the development of the required fine-grained structure with a pertinent combination of superplastic and conventional properties. Belk (163) gives detailed information which shows that a sound knowledge of the effects on the time-temperature-transformation diagram of alloying additions and of being off eutectoid composition, is necessary before a suitable processing route for a particular alloy can be developed. Casting, homogenisation, quenching, working prior to transformation and post-transformation thermomechanical treatments can all have an effect on both behaviour and subsequent properties. Progressive refinement of procedures with associated property assessment can be used to define an optimal production route. For example, it is important that macrosegregation during casting is avoided. At the other extreme, the total rolling reduction during processing and the final rolling temperature influence the resultant combination of properties. Specific details of optimised schedules are difficult to obtain although a certain amount of information is often incorporated in patent specifications.

This family of alloys has an established commercial place which is perhaps fitting in view of their important contribution to the awakening of interest in superplasticity and to the understanding of basic phenomena.

Aluminium alloys

Early investigations of superplasticity involved much work with aluminium-based alloys, especially the Al-$CuAl_2$ eutectic alloy (100,241,242). However, this and similar alloys exhibited limited room temperature ductility and subsequently commercial work was concentrated on the production of an alloy with a low (total) content of alloying additions (preferably less than 10 wt. percent) and with a stable structure as similar to conventional structures as possible but still able to exhibit superplasticity. The most worthwhile success was achieved with an aluminium -6 wt. percent copper -0.5 wt. percent zirconium alloy (245,247) subsequently marketed as Supral (see, for example, Sawle (589)). The balance of evidence supports the view (245, 247) that this alloy depends for its success on important contributions from each of the major alloying constituents. Zirconium, present as a fine dispersion of $ZrAl_3$, plays a vital role in inhibiting recrystallisation and restricting grain growth. It is essential that fast cooling rates are used during the solidification stage to retain the zirconium in supersaturated solid solution and to minimise segregation. Only then can the required fine dispersion of $ZrAl_3$ be developed by heat-treatment. Copper helps also to control recrystallisation during deformation and heat-treatment such that a recovered structure is largely present prior to superplastic forming. During superplastic forming an ultra-fine grain structure develops.

This alloy is now being produced as sheet on a large scale and a wide range of sheet-thermoformed parts have been manufactured. Some typical examples are shown in Fig.6.22. Attention has also been given

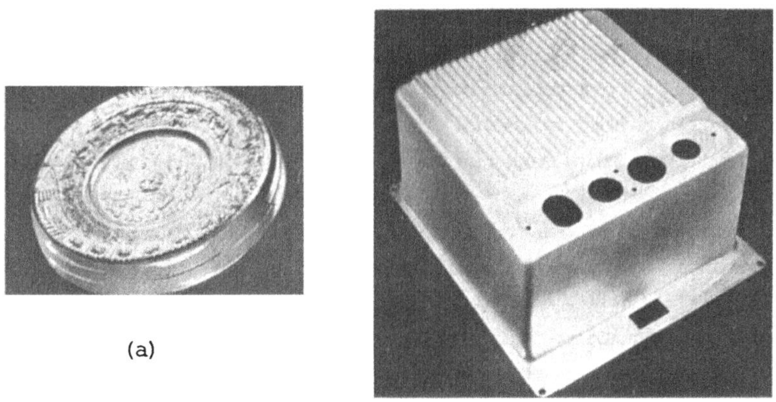

(a) (b)

Figure 6.22 Parts manufactured by superplastic forming of Supral (a) decorative panel, (b) computer housing (courtesy TI Superform Limited)

to improving corrosion resistance and rendering the alloy suitable for decorative anodising treatments by cladding the alloy with thin surface layers of high-purity aluminium (245).

No details are available of the service fatigue and creep behaviour and it is possible that the alloy will be prone to cavitation because of the presence of a significant volume fraction of $CuAl_2$ particles. Nevertheless if this was the case it should be an easily solved problem since it has been established that satisfactory microstructural and deformation characteristics have been exhibited by related alloys (247), e.g. Al-Zn-Zr and Al-Zn-Mg-Zr, in which all precipitates (other than $ZrAl_3$) are in solution during the critical stages.

The future for the aluminium-based superplastic alloys is highly promising.

Stainless steels

Commercial application of superplasticity in steels to yield enhanced formability has not met with success. To date the nearest approach has been the development (277) of a "microduplex" iron-26% chromium-6.5% nickel stainless steel known as IN744X. By careful control of carbon content (to $<$ 0.03%C) the two-phase austenite-ferrite structure was enabled to exhibit good hot workability with superplastic extensions of several hundred percent. It was also capable of being cold-worked to a variety of forms. Furthermore, the fine duplex structure showed superior mechanical properties (tensile and fatigue) and improved corrosion resistance (except in boiling $MgCl_2$ solution) when compared to conventional stainless steels. Minor variations in composition did not appear to be critical and the alloy could easily be welded using conventional methods. When the welds had to be introduced prior to forming, a post-welding heat treatment ensured their satisfactory performance during subsequent superplastic forming. A critical step that was overcome was scaling up the laboratory-scale manufacturing practice to the production of industrial ingots. Clearly this class of alloys has promise and it is to be hoped that further evidence of their utilisation in superplastic forming is soon forthcoming, since so far only the application of tube expansion into a ceramic mould has been demonstrated (559).

Titanium alloys

Although many titanium alloys are known to be superplastic (see also Table 3.1) most practical use has been made of the titanium-6%

aluminium-4% vanadium alloy first studied by Lee and Backofen (284). The superplastic capability of this alloy is strongly affected by its metallurgical condition. Optimal superplastic response was produced by heavily hot-working in the ($\alpha+\beta$) phase field to yield a fine-grained structure (~3μm). Insufficient hot reduction led to larger grain sizes which markedly reduced the superplastic formability. When pre-welded (say, by the TIG welding process) sheets were used, the behaviour of the welds during superplastic deformation was satisfactory (304). Pressure thermoforming of this alloy with hot argon (590) has been shown to be a feasible route for fabricating a range of parts for high-performance aircraft including intake ducts, engine nacelle beam frames etc. (Fig.6.23). It is of interest to note that it has been reported that cost and weight studies indicate more than a 30% weight saving and more than a 50% cost saving for the superplastically formed part illustrated. The same process has also been used for the production of vacuum chambers and fuel tanks (591). After superplastic forming the alloy can be heat-treated above the β-transus to restore acceptable room temperature properties.

Nickel-base superalloys

The utilisation of superplasticity in nickel-base superalloys was considered in Section 6.1.9 when the powder metallurgy process known as "Gatorising" was referred to. Gatorising utilises the low flow stresses at high temperatures of the compacted alloy powders to produce forgings with extremely close tolerances. In so doing a decrease in charge weight and a lowering of machining costs results. There is an additional cost saving because of the low forging pressures. These require the use of only light forging equipment. A number of alloys have been successfully forged including IN-100 (12.5Cr, 18.3Co,3.2Mo,4.4Ti,4.9Al), Astroloy (15Cr, 15Co,5.25Mo,3.5Ti,4.4Al) and Waspalloy (19.5Cr,13.5Co,4.3Mo, 3.0Ti,1.4Al) (283,576,592). Most effort has been devoted to the manufacture of turbine discs (Fig.6.24). In practice this is one of the very few methods available for easily forming parts from these highly creep resistant alloys. After forming the final properties are uniform and excellent and the product exhibits a high degree of creep resistance at the working temperature.

There are other minor examples of potentially useful alloys which are amenable to superplastic forming (see, for instance, Edington

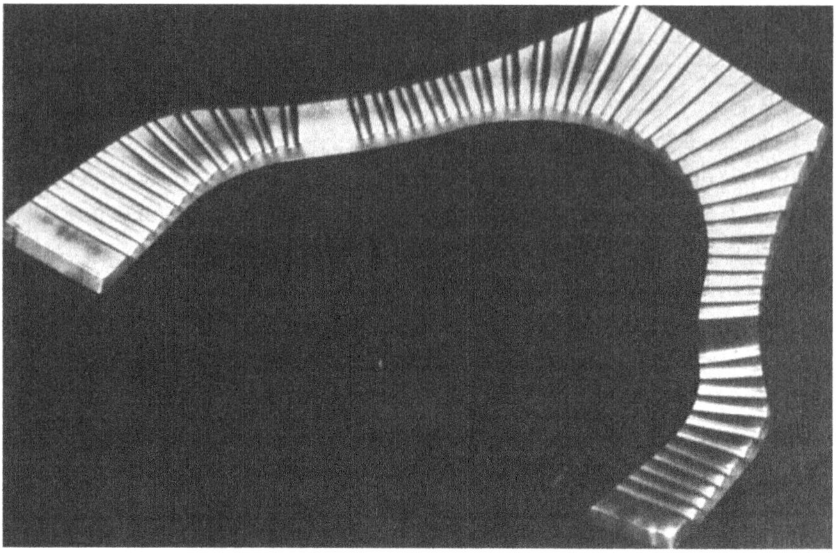

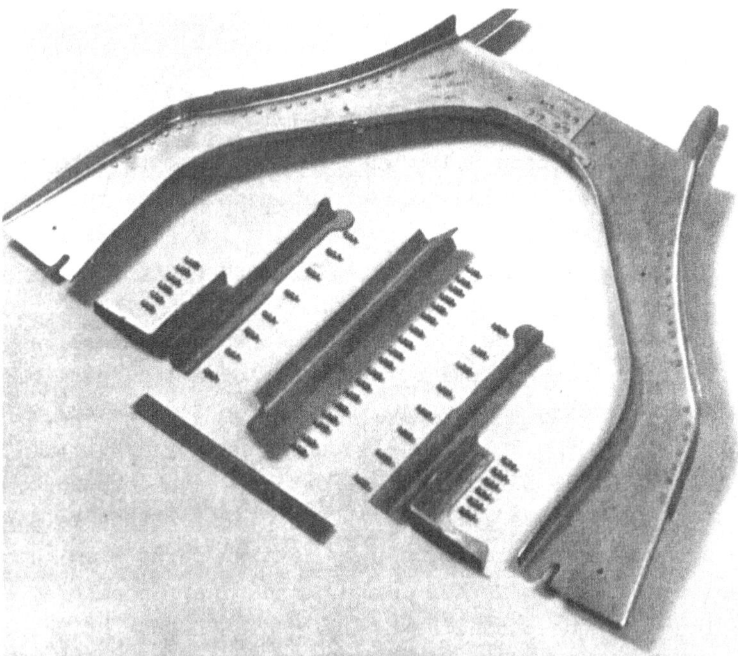

Figure 6.23 (a) Superplastically formed nacelle beam frame made with all fittings in one piece from Ti-6Al-4V alloy, (b) Conventionally assembled the frame is made up of 104 components (courtesy Rockwell International)

Figure 6.24 Turbine disc forgings produced by the gatorising process (courtesy Pratt and Whitney Aircraft)

(80)) but they still await development on a commercial scale.

6.3.2 Design considerations

The choice of a forming process depends on a variety of factors (554),

(i) the engineering properties required of the material;

(ii) the type of shape to be produced;

(iii) quantity of parts required;

(iv) extent of secondary finishing operations;

(v) tooling costs, and

(vi) the equipment available.

In addition, it is necessary to choose between sheet forming or a bulk-forming process. The former is most likely to involve the production of parts with comparatively deep sections and complex, compound curves. The latter process is used mainly for bulky items such as machine components.

Sheet forming

For the design of sheet formed parts, design guidelines are available from manufacturers. The most important parameters are depth of draw, draft angles and corner radii. For forming with Supral sheets (589) the limiting aspect ratio depends on the process used and a maximum height:width ratio of 0.6 is recommended for male forming while with female forming this reduces to 0.4. The draft angles are required to be not less than 2 degrees and 3 degrees for male and female forms, respectively. There are also constraints imposed on corner radii. The various design parameters are shown in Fig.6.25.

In manufacturing parts it is important to make use of the ability to form in sub-configurations such as strengthening ribs. Another important consideration is the forming ratio specified as the ratio of the surface area of the formed part to that of the starting material. In order to avoid thinning this is commonly kept below $\sim$4:1 although much latitude exists because of the control inherent in the choice of forming procedure.

Bulk forming

Here there are effectively no constraints since large strains can readily be accommodated and superplastic processes can be used to produce extremely fine detail.

6.3.3 Tooling

As discussed by Young et al. (592) one of the strong incentives for making use of superplastic forming as a production process is the ability to make use of simple tooling. The geometrical form of tooling is closely similar to that used in conventional operations. However, even for relatively rapid operations, the low forming stresses make it possible to use light, inexpensive materials without the strength required of normal forging or pressing die assemblies. Furthermore, with sheet forming it is frequently possible to gain a further cost advantage by using only one half of mould-die set. Preferably this has a good surface finish since the nature of superplastic forming is such that any surface irregularities on the die face are faithfully repro-

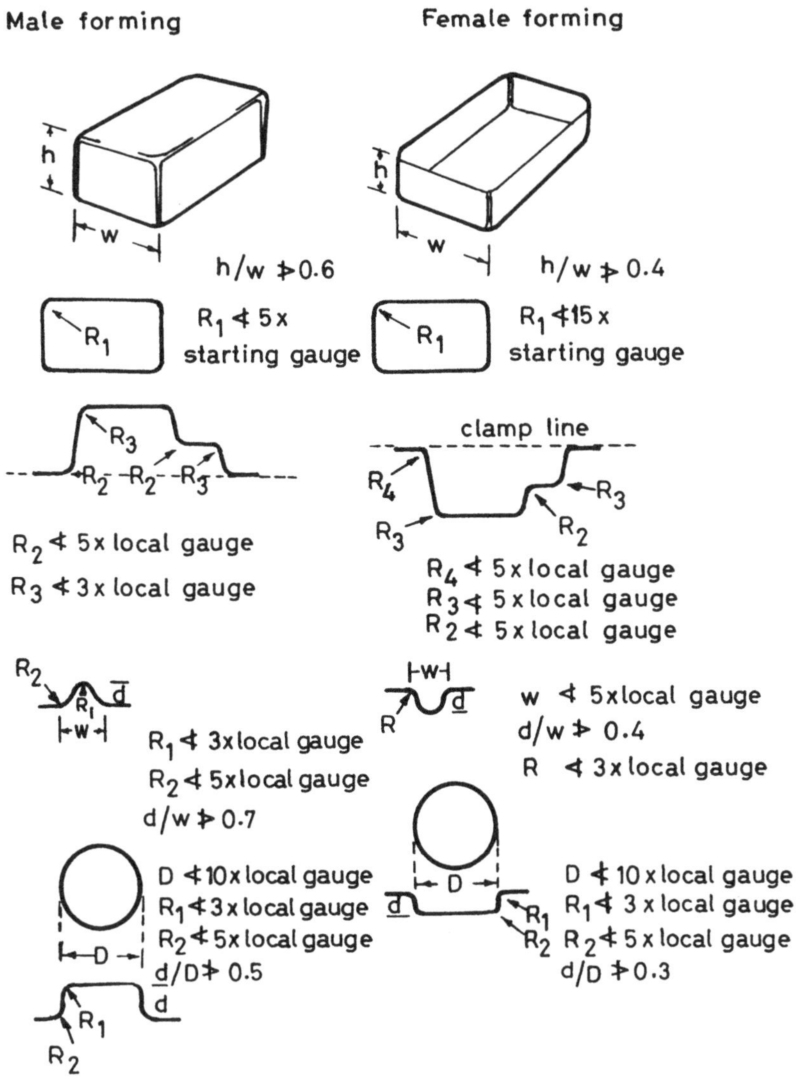

Figure 6.25 Typical design parameters to be used when forming with Supral sheet (courtesy T.I.Superform Limited)

duced. The tools also require venting to avoid air entrapment and lubrication to facilitate release and part removal. In most cases the use of pin ejectors is not necessary.

Because of the low forming pressures and temperatures a wide range of die materials has been used for superplastic forming. For low numbers of a product simple ceramic, concrete or plaster of paris moulds are suitable. Superior performance is obtained from graphite moulds (these are self-lubricating but dirty). The longest production runs are possible when working with cast or fabricated metal moulds. Nevertheless, in these cases it is seldom necessary to use high-performance alloys since quite acceptable results are obtainable with aluminium or mild steel.

In some cases with large closed-die forgings and/or the higher temperature materials (titanium- or nickel-base alloys) the use of more conventional tool and die steels becomes imperative.

In general, however, the problems encountered in tool design and construction are minimal in comparison with those associated with the development of dies for conventional forming operations.

References

1. Davies, G.J.; Edington, J.W.; Cutler, C.P.; Padmanabhan, K.A.: J. Mater. Sci. 5 (1970) 1091

2. Rosenhain, W.; Haughton, J.L.; Bingham, K.E.: J. Inst. Met. 23 (1920) 261

3. Bochvar, A.A.; Sviderskaya, Z.A.: Izvest. Akad. Nauk SSSR Otdel Tekhn. Nauk 9 (1945) 821; see also Metall.Abstr. 13 (1946) 318; Chem. Abstr. 41 (1947) 2375

4. Underwood, E.E.: J. Met. 14 (1962) 914

5. Hanson, D.; Gayler, M.L.V.: J. Inst. Met. 34 (1925) 125

6. Jeffries, Z.; Archer, R.S.: The Science of Metals. New York: McGraw Hill 1924, p.76

7. Hargreaves, F.: J. Inst. Met. 38 (1927) 315

8. Hargreaves, F.: J. Inst. Met. 39 (1928) 301

9. Hargreaves, F.: Hills, R.J.: J. Inst. Met. 40 (1928) 41

10. Hargreaves, F.; Hills, R.J.: J. Inst. Met. 41 (1929) 257

11. Jenkins, C.M.H.: J. Inst. Met. 40 (1928) 41

12. Pearson, C.E.: J. Inst. Met. 54 (1934) 111

13. Fell, E.W.: J. Inst. Met. 54 (1934) 125

14. J. Inst. Met. 54 (1934) pp. 131 et.seq.

15. Cook, G.: J. Inst. Met. 54 (1934) 134

16. Haigh, B.P.: J. Inst. Met. 54 (1934) 137

17. Sauveur, A.: Trans. AIMME 70 (1924) 3

18. Sauveur, A.: Iron Age 113 (1924) 581

19. Clark, C.L.; Russ, J.J.: Trans. AIMME 167 (1946) 736

20. Thrig, H.K.: Trans. AIMME 167 (1946) 749

21. Scheil, E.: Z. Anorg. Chem. 207 (1932) 21

22. Wassermann, G.: Arch. Eisenhuett 6 (1933) 347

23. Wassermann, G.: Arch. Eisenhuett 10 (1937) 321

24. Scheil, E.; Theile, W.: Arch. Eisenhuett 10 (1937) 477

25. Wassermann, G.: Arch. Eisenhuett 11 (1937) 89

26. Hornbogen, E.; Wassermann, G.: Z. Metallk. 47 (1956) 427

27. de Jong, M.; Rathenau, G.W.: Acta Metall. 7 (1959) 246

28. Lozinsky, M.G.; Simeonova, I.S.: Acta Metall. 7 (1959) 709

29. Oelschlaegel, D.; Weiss, V.: Trans. Am. Soc. Met. 59 (1966) 143

30. Bochvar, A.A.: Izvest. Akad. Nauk SSR (Tekhn.) (5) (1946) 743; see also Chem. Abstr. 41 (1947) 1190

31. Bochvar, A.A.; Sviderskaya, Z.A.: Izvest. Akad. Nauk SSR (Tekhn.) (7) (1946) 1001; see also Metall. Abstr. 14 (1947) 96; Chem. Abstr. 41 (1947) 1589

32. Bochvar, A.A.: Izvest. Akad. Nauk SSSR Otdel Tekhn. Nauk 5 (1948) 649; see also Chem. Abstr. 45 (1951) 7501

33. Chaudhari, P.: Acta Metall. 15 (1967) 1777

34. Presnyakov, A.A.; Chervyakova, V.V.: Izvest. Akad. Nauk SSR (Tekhn.) (3) (1958) 120; see also Metall. Abstr. 26 (1958-59) 237; Chem. Abstr. 53 (1959) 2045

35. Presnyakov, A.A.; Chervyakova, V.V.: Vestnik. Akad. Nauk Kazakh. SSR 14 (12) (1958) 76; see also Chem. Abstr. 53 (1959) 7929

36. Presnyakov, A.A.; Starikova, G.V.: Izvest. Akad. Nauk SSSR Met. i. Topl. (Tekhn.) (1) (1959) 75; see also Metall. Abstr. 27 (1959-60) 93

37. Presnyakov, A.A.; Starikova, G.V.; Samoilov, V.A.; Chervyakova, V.V.: Izvest. Akad. Nauk SSSR Met. i. Topl. (Tekhn.) (2) (1961) 146; see also Metall. Abstr. 29 (1961-62) 135; Chem. Abstr. 55 (1961) 1966

38. Presnyakov, A.A.; Chervyakova, V.V.: Phys. Met. Metallog. 8 (1) (1959) 96

39. Presnyakov, A.A.; Chervyakova, V.V.: Russ. Metallurgy and Fuels (3) (1960) 85

40. Starikova, G.V.; Presnyakov, A.A.: Phys. Met. Metallog. 13 (5) (1962) 125

41. Presnyakov, A.A.; Starikova, G.V.: Russ. Metallurgy and Mining (5) (1963) 95

42. Backofen, W.A.; Turner, I.R.; Avery, D.H.: Trans. Am. Soc. Met. 57 (1964) 980

43. Dieter, G.E.: Mechanical Metallurgy. New York: McGraw Hill 1961, p. 237

44. Ludwick, P.: Elemente der technologischen Mechanik. Berlin: Springer Verlag 1909

45. Zener, C.; Hollomon, J.H.: J. Appl. Phys. 17 (1946) 69

46. Hollomon, J.H.: Trans. AIMME 171 (1947) 535

47. Dorn, J.E.; Goldberg, A.; Tietz, T.E.: Trans. AIMME 180 (1949) 205

48. Hart, E.W.: Acta Metall. 18 (1970) 599

49. Appleby, E.J.: J. Appl. Phys. 41 (1970) 4902

50. Wray, P.J.; Richmond, O.: J. Appl. Phys. 39 (1968) 5754

51. Rossard, C.: Rev. Metall. 63 (1966) 225

52. Hart, E.W.: Acta Metall. 15 (1967) 351

53. Campbell, J.D.: J. Mech. Phys. Sol. 15 (1967) 359

54. Stuewe, H.: Z. Metallk. 61 (1970) 704

55. Wray, P.J.: J. Appl. Phys. 41 (1970) 3347

56. Dunlop, G.L.; Taplin, D.M.R.: J. Mater. Sci. 7 (1972) 84

57. Lubahn, J.D.: Trans. ASME 68 (1946) 277

58. Nadai, A.; Manjoine, M.J.: J. Appl. Mech. 8 (1941) A77

59. Orowan, E.: Rep. Progr. Phys. 12 (1949) 185

60. Rai, G.; Grant, N.J.: Metall. Trans. 6A (1975) 385

61. Hazzledine, P.M.; Newbury, D.E.: Grain Boundary Structure and Properties (eds. Chadwick, G.A.; Smith, D.A.). London: Academic Press 1976, p. 235

62. Ashby, M.F.; Verrall, R.A.: Acta Metall. 21 (1973) 149

63. Padmanabhan, K.A.: Mater. Sci. Eng. 29 (1977) 1

64. Jovane, F.: Int. J. Mech. Sci. 10 (1968) 403

65. Al-Naib, T.Y.M.; Duncan, J.L.: Int. J. Mech. Sci. 12 (1970) 463

66. Ghosh, A.; Duncan, J.L.: Int. J. Mech. Sci. 12 (1970) 499

67. Hill, R.: The Mathematical Theory of Plasticity. Oxford: Clarendon Press 1950, pp. 26, 30

68. Nicholson, R.B.: Electron Microscopy and the structure of materials (ed. Thomas, G.). Berkeley: University of California Press 1972, p. 689

69. Padmanabhan, K.A.; Davies, G.J.: Phys. Stat. Solidi (a) 18 (1973) 295

70. Avery, D.H.; Backofen, W.A.: Trans. Am. Soc. Met. 58 (1965) 551

71. Pompe, W.; Feige, B.: Phys. Stat. Solidi (a) 5 (1971) 359

72. Padmanabhan, K.A.; Davies, G.J.: Rheol. Acta 13 (1974) 434

73. Suery, M.; Baudelet, B.: J. Mater. Sci. 10 (1975) 1022

74. Padmanabhan, K.A.; Moles, M.D.C.; Davies, G.J.: J. Mater. Sci. 12 (1977) 632

75. Alden, T.H.: Trans. Met. Soc. AIME 236 (1966) 1633

76. Alden, T.H.: Acta Metall. 15 (1967) 469

77. Hayden, H.W.; Gibson, R.C.; Merrick, H.F.; Brophy, J.H.: Trans. Am. Soc. Met. 60 (1967) 3

78. Nuttall, K.; Nicholson, R.B.: Phil. Mag. 17 (1968) 1087

79. Ball, A.; Hutchison, M.M.: Met. Sci. J. 3 (1969) 1

80. Edington, J.W.: Met. Technol. 3 (1976) 138

81. Sagat, S.; Taplin, D.M.R.: Met. Sci. 10 (1976) 94

82. Suery, M.; Baudelet, B.: J. Mater. Sci. 8 (1973) 363

83. Reznikov, V.G.: Phys. Met. Metallog. 36 (6) (1973) 190

84. Arieli, A.; Rosen, A.: Scripta Metall. 10 (1976) 809

85. Naziri, H.; Pearce, R.: J. Inst. Met. 97 (1969) 326

86. Cahoon, J.R.: Met. Sci. 9 (1975) 346

87. Fields Jr., D.S.; Stewart, T.J.: Int. J. Mech. Sci. 13 (1971) 63

88. Cottingham, D.M.; Watts, B.M.: Appendix to Watts, B.M.; Stowell, M.J.: J. Mater. Sci. 6 (1971) 228

89. Newbury, D.E.; Joy, D.C.: Scripta Metall. 4 (1970) 825

90. Baudelet, B.; Suery, M.: J. Mater. Sci. 7 (1972) 512

91. Hayden, H.W.; Brophy, J.H.: Trans. Am. Soc. Met. 61 (1968) 542

92. Hedworth, J.; Stowell, M.J.: J. Mater. Sci. 6 (1971) 1061

93. Herriot, G.; Suery, M.; Baudelet, B.: Scripta Metall. 6 (1972) 657

94. Clark, M.A.; Alden, T.H.: Acta Metall. 21 (1973) 1195

95. Ducheyne, P.; De Meester, P.: J. Mater. Sci. 9 (1974) 109

96. Moles, M.D.C.; Davies, G.J.: Scripta Metall. 10 (1976) 455

97. Sagat, S.; Taplin, D.M.R.: Acta Metall. 24 (1976) 307

98. Newbury, D.E.: J. Mater. Sci. 8 (1973) 145

99. Padmanabhan, K.A.; Davies, G.J.: Met. Sci. 11 (1977) 177

100. Holt, D.L.; Backofen, W.A.: Trans. Am. Soc. Met. 59 (1966) 755

101. Morrison, W.B.: Trans. Am. Soc. Met. 61 (1968) 423

102. Smith, C.I.; Ridley, N.: Met. Technol. 1 (1974) 191

103. Smith, C.I.; Norgate, B.; Ridley, N.: Met. Sci. 10 (1976) 182

104. Naziri, H.; Pearce, R.: J. Inst. Met. 101 (1973) 197

105. Padmanabhan,K.A.: Trans. Ind. Inst. Met. 26 (3) (1973) 41

106. Klepaczko, J.: Int. J. Mech. Sci.10 (1968) 297

107. Jonas, J.J.; Holt, R.A.; Coleman, C.E.: Acta Metall. 24 (1976) 911

108. Argon, A.S.: The Inhomogeneity of Plastic Deformation. Metals Park Ohio: Am. Soc. Metals 1973, Chap. 7

109. Hutchinson, J.W.; Obrecht, H.: Fracture 1977 Vol. 1 (ed. Taplin, D.M.R.). Ontario: Univ. Waterloo Press 1977, p. 101

110. Tonejc, A.; Poirier, J.P.: Scripta Metall. 9 (1975) 555

111. Garde, A.M.; Chung, H.M.; Kaussner, T.F.: Acta Metall. 26 (1978) 153

112. Hutchinson, J.W.; Neale, K.W.: Acta Metall. 25 (1977) 839

113. Troost, A.; El-Magd, E.: Mater. Sci. Eng. 28 (1977) 201

114. Ghosh, A.K.; Ayres, R.A.: Metall. Trans. 7A (1976) 1589

115. Ayres, R.A.: Metall. Trans. 8A (1977) 487

116. Ghosh, A.K.: Acta Metall. 25 (1977) 1413

117. Langdon, T.G.: Scripta Metall. 11 (1977) 997

118. Hoff, N.J.: J. Appl. Mech. 20 (1953) 105

119. Langdon, T.G.; Taplin, D.M.R.: S.M.Arch. 2 (1977) 329

120. Avery, D.H.; Stuart, J.M.: Surfaces and Interfaces II (eds. Burke, J.J.; Reed, N.L.; Weiss, V.). New York: Syracuse Univ. Press 1968, p. 371

121. Holt, D.L.: Ultrafine grain Metals (eds. Burke, J.J.; Weiss, V.). New York: Syracuse Univ. Press 1970, p. 355

122. Hirano, T.; Iharagi, M.; Yamaguchi, M.; Yamane, T.: Metall. Trans. 6A (1975) 2159

123. Nuttall, K.: Scripta Metall. 10 (1976) 835

124. Langdon, T.G.: Proc. V New Zealand Science of Materials Conf. Wellington 1977, p. 11

125. Woodford, D.A.: Trans. Am. Soc. Met. 62 (1969) 291

126. Stewart, M.J.: Metall. Trans. 7A (1976) 399

127. Mohamed, F.A.; Ahmed, M.M.I.; Langdon, T.G.: Metall. Trans. 8A (1977) 933

128. Piatti, G.; Pellegrini, G.; Trippodo, R.: J. Mater. Sci. 11 (1976) 186

129. Edington, J.W.; Melton, K.N.; Cutler, C.P.: Progr. Mater. Sci. 21 (1976) 61

130. Morrison, W.B.: Trans. Met. Soc. AIME 242 (1968) 2221

131. Kaibyshev, O.A.; Kazachkov, I.V.; Rodionov, B.V.: Phys. Met. Metallog. 39 (2) (1975) 102

132. Padmanabhan, K.A.: Ph.D. Thesis, Univ. Cambridge 1971

133. Arieli, A.; Rosen, A.: Scripta Metall. 10 (1976) 471

134. Hildebrand, H.; Michalzik, G.; Simmon, B.: Met. Technol. 4 (1977) 32

135. Lequear, H.A.; Lubahn, J.D.: J. Met. 8 (1956) 497

136. Lee, D.: Metall. Trans. 1 (1970) 1607

137. Nuttall, K.: Int. J.Mech. Sci.13 (1971) 83

138. Griffiths, P.; Hammond, C.: Acta Metall. 20 (1972) 935

139. Feltham, P.: J. Inst. Met. 89 (1960) 210

140. Guiu, F.; Pratt, P.L.: Phys. Stat. Solidi 6 (1964) 111

141. Lee, D.; Hart, E.W.: Metall. Trans. 2 (1971) 1245

142. Guthrie, A.M.S.; Newbury, D.E.; Hazzledine, P.M.: Scripta Metall. 6 (1972) 841

143. Murty, G.S.: J. Mater. Sci. 8 (1973) 611

144. Herriot, G.; Baudelet, B.; Jonas, J.J.: Acta Metall. 24 (1976) 687

145. Geckinli, A.E.; Barrett, C.R.: Scripta Metall. 8 (1974) 115

146. Geckinli, A.E.; Barrett, C.R.: J. Mater. Sci. 11 (1976) 510

147. Woodford, D.A.: Metall. Trans. 7A (1976) 1244

148. Kot, R.A.; Weiss, V.: Metall. Trans. 1 (1970) 2685

149. de Jong, M.; Rathenau, G.W.: Acta Metall. 9 (1961) 714

150. Greenwood, G.W.; Johnson, R.H.: Proc. Roy. Soc. A283 (1965) 403

151. Yoder, G.R.; Weiss, V.: Metall. Trans. 3 (1972) 675

152. Murty, G.S.; Singh, U.P.; Iyengar, N.G.R.: J. Inst. Met. 101 (1973) 337

153. Marcantonio, J.A.: Metals and Materials 7 (1973) 449

154. Belk, J.A.: Metals and Materials 8 (1974) 414

155. Gibbs, G.B.: Phil. Mag. 13 (1966) 317

156. Lloyd, D.J.; Embury, J.D.: Met. Sci. J. 4 (1970) 6

157. Solomon, A.A.; Nix, W.D.: Acta Metall. 18 (1970) 863

158. Murty, G.S.: Scripta Metall. 6 (1972) 663

159. Stevens, R.N.: Metall. Reviews 11 (1966) 129

160. Holt, D.L: Trans. Met. Soc. AIME 242 (1968) 25

161. Vatsava, R.B.; Langdon, T.G.: Acta Metall. 27 (1979) 251

162. Nuttall, K.: J. Inst. Met. 99 (1971) 266

163. Belk, J.A.: Met. Technol. 3 (1976) 161

164. Garmong, G.; Paton, N.E.; Chesnutt, J.C.; Nevarez, L.F.: Metall. Trans. 8A (1977) 2026

165. Dasarathy, C.: Z. Metallk. 62 (1971) 612

166. Alden, T.H.: Trans. Am. Soc. Met. 61 (1968) 559

167. Pearce, R.; Swanson, C.J.: Sheet Metal Industries 47 (1970) 599

168. Srinivasa Rao, S.; Sivakesavan, O.; Ghude, S.H.; Tamhankar,R.V.: Trans. Ind. Inst. Met. 23 (4) (1970) 44

169. Cline, H.E.; Alden, T.H.: Trans. Met. Soc. AIME 239 (1967) 710

170. Gifkins, R.C.: J. Inst. Met. 95 (1967) 373

171. Darekar, V.S.; Chaudhari, R.D.: Trans. Ind. Inst. Met. 23(3) (1970) 56

172. Martin, P.J.; Backofen, W.A.: Trans. Am. Soc. Met. 60 (1967) 352

173. Zehr, S.W.; Backofen, W.A.: Trans. Am. Soc. Met. 61 (1968) 300

174. Cutler, C.P.; Edington, J.W.: Met. Sci. J. 5 (1971) 201

175. Burton, B.: Scripta Metall. 5 (1971) 669

176. Rawal, S.P.; Murty, G.S.: Trans. Jap. Inst. Met. 13 (1972) 57

177. Murty, G.S.: Scripta Metall. 6 (1972) 1075

178. Murty, G.S.: J. Mater. Sci. 8 (1973) 611

179. Horiuchi, R.; El-Sebai, A.B.; Otsuka, M.: Phys. Stat. Solidi(a) 21 (1974) K89

180. Ahmed, M.M.I.; Langdon, T.G.: Metall. Trans. 8A (1977) 1832

181. Agarwal, S.B.; Vaidya, M.L.: Scripta Metall. 9 (1975) 447

182. Prematta, R.J.; Venkatesan, P.S.; Pense, A.: Metall. Trans. 7A (1976) 1235

183. Moles, M.D.C.; Davies, G.J.: Met. Sci. 10 (1976) 314

184. Williams, D.A.C.; Naziri, H.; Pearce, R.: Scripta Metall. 3 (1969) 117

185. Edwards, G.R.; Shyne, J.C.; Sherby, O.D.: Metall. Trans. 2 (1971) 2955

186. Cook, R.C.; Risebrough, N.R.: Scripta Metall. 2 (1968) 487

187. Naziri, H.; Pearce, R.: Scripta Metall. 3 (1969) 807

188. Naziri, H.; Pearce, R.: Scripta Metall. 3 (1969) 811

189. Naziri, H.; Pearce, R.: J. Inst. Met. 98 (1970) 71

190. Naziri, H.; Pearce, R.: Acta Metall. 22 (1974) 1321

191. Valiev, R.Z.; Kaibyshev, O.A.: Phys. Met. Metallog. 41(2) (1976) 137

192. Johnson, R.H.; Packer, C.M.; Sherby, O.D.: Trans. Am. Soc. Met. 60 (1967) 564

193. Johnson, R.H.; Packer, C.M.; Anderson, L.; Sherby, O.D.: Phil. Mag. 18 (1968) 1309

194. Packer, C.M.; Johnson, R.H.; Sherby, O.D.: Trans. Met. Soc. AIME 242 (1968) 2485

195. Marshall, J.C.; Stewart, T.J.; Babcock, T.C.: Met. Eng. Quart. 13(4) (1973) 12

196. Alden, T.H.; Schadler, H.W.: Trans. Met. Soc. AIME 242 (1968) 825

197. Kossowsky, R.; Bechtold, J.H.: Trans. Met. Soc. AIME 242 (1968) 716

198. Holt, D.L.: Trans. Met. Soc. AIME 242 (1968) 740

199. Fike, K.D.; Rack, H.J.: Trans. Am.Soc. Met. 62 (1969) 537

200. Barlow, I.C.: J. Mater. Sci. 4 (1969) 835

201. Lee, E.U.; Underwood, E.E.: Met. Trans. 1 (1970) 1399

202. Presnyakov, A.A.; Azhibayeva, A.S.M.; Soymin, N.Ya.; Aubakirova, R.K.; Karpenyuk, A.N.: Phys. Met. Metallog. 32(3) (1971) 214

203. Lee, D.: J. Inst. Met. 99 (1971) 66

204. Nuttall, K.: J. Inst. Met. 99 (1971) 291

205. Kaibyshev, O.A.; Kazachkov, I.V.: Phys. Met. Metallog. 34(2) (1972) 163

206. Biswas, C.R.D.; Murty, G.S.: Trans. Jap. Inst. Met. 13 (1972) 8

207. Nuttall, K.: J. Inst. Met. 100 (1972) 114

208. Naziri, H.; Pearce, R.; Brown, M.H.; Hale, K.F.: J.Microscopy 97 (1973) 229

209. Vaidya, M.L.; Lingamurty, K.; Dorn, J.E.: Acta Metall. 21 (1973) 1615

210. Melton, K.N.; Edington, J.W.: Scripta Metall. 8 (1974) 1141

211. Naziri, H.; Pearce, R.; Brown, M.H.; Hale, K.F.: Acta Metall. 23 (1975) 489

212. Ishikawa, H.; Mohamed, F.A.; Langdon, T.G.: Phil. Mag. 32 (1975) 1269

213. Holt, D.L.: Trans. Am. Soc. Met. 60 (1967) 564

214. Melton, K.N.; Edington, J.W.: J. Inst. Met. 101 (1973) 122

215. Melton, K.N.; Edington, J.W.; Kallend, J.S.; Cutler, C.P.: Acta Metall. 22 (1974) 165

216. Melton, K.N.; Edington, J.W.: Met. Sci. J. 7 (1973) 172

217. Melton, K.N.; Edington, J.W.: Scripta Metall. 9 (1975) 559

218. Naziri, H.; Pearce, R.: Int. J. Mech. Sci. 12 (1970) 513; see also comment by Cross, C.R.: Int. J.Mech. Sci. 13 (1971) 77

219. Smith, A.E.W.; Hare, G.A.: J. Inst. Met. 101 (1973) 320

220. Nuttall, K.: J. Inst. Met. 101 (1973) 329

221. Lee, J.D.; Niessen, P.: Metall. Trans. 4 (1973) 949

222. Lee, J.D.; Niessen, P.: J. Mater. Sci. 9 (1974) 1467

223. McCarthy, W.H.; Shyne, J.C.; Sherby, O.D.: Nature 208 (1965) 579

224. Tromans, D.; Lund, J.A.: Trans. Am. Soc. Met. 59 (1966) 672

225. Merz, M.D.: J. Nucl. Mater. 41 (1971) 348

226. Merz, M.D.; Nelson, R.D.: Metall. Trans. 2 (1971) 1399

227. Dahlgren, S.D.: Trans. Met. Soc. AIME 242 (1968) 126

228. Allen, R.P.; Arrowsmith, H.W.; Nelson, R.D.: Metall. Trans. 2 (1971) 344

229. Chapman, J.A.; Wilson, D.V.: J. Inst. Met. 91 (1962) 39

230. Lee, D.: Acta Metall. 17 (1969) 1057

231. Lee, D.: Metall. Trans. 1 (1970) 309

232. Greenfield, P.; Vickers, W.: J. Nucl. Mater. 22 (1967) 77

233. Vickers, W.; Greenfield, P.: J. Nucl. Mater. 24 (1967) 249

234. Woodford, D.A.: J. Inst. Met. 96 (1968) 371

235. Backofen, W.A.; Azzarto, F.J.; Murty, G.S.; Zehr, S.W.: Ductility. Metals Park Ohio: Am. Soc. Metals 1968, p. 279

236. Karim, A.; Backofen, W.A.: Mater. Sci. Eng.3 (1968-69) 306

237. Backofen, W.A.; Murty, G.S.; Zehr, S.W.: Trans. Met. Soc. AIME 242 (1968) 329

238. Karim, A.; Holt, D.L.; Backofen, W.A.: Trans. Met. Soc. AIME 245 (1969) 1131

239. Likhachev, V.A.; Myshlyaev, M.M.; Olevskii, S.S.; Chuchman, T.N.: Acta Metall. 22 (1974) 829

240. Beghi, G.; Matera, R.; Piatti, G.: J. Mater. Sci. 5 (1970) 820

241. Stowell, M.J.; Robertson, J.L.; Watts, B.M.: Met. Sci. J. 3 (1969) 41

242. Watts, B.M.; Stowell, M.J.: J. Mater. Sci. 6 (1971) 228

243. Schmidt-Whitley, R.D.: Z. Metallk. 64 (1973) 552

244. Cutler, C.P.; Edington, J.W.; Kallend, J.S.; Melton, K.N.: Acta Metall. 22 (1974) 665

245. Grimes, R.; Baker, C.; Stowell, M.J.; Watts, B.M.: Aluminium 51 (1975) 720

246. Watts, B.M.; Stowell, M.J.; Baikie, B.L.; Owen, D.G.E.: Met. Sci. 10 (1976) 189

247. Grimes, R.; Stowell, M.J.; Watts, B.M.: Met. Technol. 3 (1976) 154

248. Horiuchi, R.; El-Sebai, A.B.; Otsuka, M.: Scripta Metall. 7 (1973) 1101

249. Matsuki, K.; Yamada, M.: J. Jap. Inst. Met. 37 (1973) 448

250. Towler, B.; Cutler, C.P.; Edington, J.W.: J. Inst. Met. 101 (1973) 332

251. Bricknell, R.H.; Edington, J.W.: Metall. Trans. 7A (1976) 153

252. Tavassoli, A.A.; Razavi, S.E.; Fallah, N.M.: Metall. Trans. 6A (1975) 591

253. Matsuki, K.; Uetani, Y.; Yamada, M.; Murakami, Y.: Met. Sci. 10 (1976) 235

254. Wilcox, B.A.; Clauer, A.H.: The Microstructure and Design of Alloys. London: Metals Soc. 1973, p.227

255. Mobley, C.E.; Clauer, A.H.; Wilcox, B.A.: J. Inst. Met. 100 (1972) 142

256. Cline, H.E.; Lee, D.: Acta Metall. 18 (1970) 315

257. Dunlop, G.L.; Taplin, D.M.R.: J. Mater. Sci 7 (1972) 316.

258. Taplin, D.M.R.; Sagat, S.: Mater. Sci. Eng. 9 (1972) 53

259. Sagat, S.; Blenkinsop, P.A.; Taplin, D.M.R.: J. Inst. Met. 100 (1972) 268

260. Dunlop, G.L.; Taplin, D.M.R.: J. Aust. Inst. Met. 16 (1971) 195

261. Dunlop, G.L.; Reid, J.D.; Taplin, D.M.R.: Metall. Trans. 2 (1971) 2308

262. Dunlop, G.L.; Shapiro, E.; Taplin, D.M.R.; Crane, J.: Metall. Trans. 4 (1973) 2039

263. Bright, M.W.A.; Taplin, D.M.R.: Copper Development Association Publication 061/2. Cleveland: Am. Soc. Met. 1972

264. Fleck, R.G.; Taplin, D.M.R.: Can. Met. Quart. 11 (1972) 299

265. Butt, S.H.: Copper Development Association Publication 072/2. Clveland: Am. Soc. Met. 1972

266. Fleck, R.G.; Taplin, D.M.R.: J. Mater. Sci. 8 (1973) 1052

267. Fleck, R.G.; Beevers, C.J.; Taplin, D.M.R.: J. Mater. Sci. 9 (1974) 1737

268. Schelleng, R.D.; Reynolds, G.H.: Metall. Trans. 4 (1973) 2199

269. Heiple, C.R.: Metall. Trans. 4 (1973) 585

270. Schadler, H.W.: Trans. Met. Soc. AIME 242 (1968) 1281

271. Marder, A.R.: Trans. Met. Soc. AIME 245 (1969) 1337

272. Sherby, O.D.; Walser, B.; Young, C.M.; Cady, E.M.: Scripta Metall. 9 (1975) 569

273. Snape, E.; Church, N.L.: J. Met. 24(1) (1972) 23

274. Hirano, T.; Yamaguchi, M.; Yamane, T.: Metall.Trans. 5 (1974) 1249

275. Stewart, M.J.: Metall. Trans. 7A (1976) 399

276. Humpries, C.W.; Ridley, N.: J. Mater. Sci. 9 (1974) 1429

277. Gibson, R.C.; Hayden, H.W.; Brophy, J.H.: Trans. Am. Soc. Met. 61 (1968) 85

278. Weinstein, D.: Trans. Met. Soc. AIME 245 (1969) 2041

279. Hayden, H.W.; Floreen, S.; Goodell, P.G.: Metall. Trans. 3 (1972) 833

280. Floreen, S.: Scripta Metall. 1 (1967) 19

281. Kaibyshev, O.A.; Markelov, A.A.: Phys. Met. Metallog. 41(1) (1976) 165

282. Lindinger, R.J.; Gibson, R.C.; Brophy, J.H.: Trans. Am. Soc. Met. 62 (1969) 230

283. Reichman, S.H.; Smythe, J.W.: Int. J. Powder Met. 6(1) (1970) 65

284. Lee, D.; Backofen, W.A.: Trans. Met. Soc. AIME 239 (1967) 1034

285. Lyttle, J.E.; Fischer, G.; Marder, A.R.: J. Met. 17 (1965) 1055

286. Arieli, A.; Rosen, A.: Metall. Trans. 8A (1977) 1591

287. Cline, H.E.: Trans. Met. Soc. AIME 239 (1967) 1906

288. Stephens, J.R.; Klopp, W.D.: Trans. Met. Soc. AIME 236 (1966) 1637

289. Garfinkle, M.; Witzke, W.R.; Klopp, W.D.: Trans. Met. Soc. AIME 245 (1969) 303

290. Garfinkle, M.; Witzke, W.R.; Klopp, W.D.: Metall. Trans. 1 (1970) 1046

291. Nicholson, R.B.: Plasticity and Superplasticity. London: Institution of Metallurgists Review Course Ser. 2(3) 1969, p. 19

292. Duncan, J.L.; Shapiro, E.; Crane, J.; Taplin, D.M.R.: Symposium on Applications of Solid Mechanics. University of Waterloo 1972

293. Mohamed, F.A.; Langdon, T.G.: Acta Metall. 23 (1975) 117

294. Mohamed, F.A.; Shei, S.; Langdon, T.G.: Acta Metall. 23 (1975) 1443

295. Langdon, T.G.; Mohamed, F.A.: Scripta Metall. 11 (1977) 575

296. Karim, A.: Scripta Metall. 3 (1969) 887

297. Karim, A.; Backofen, W.A.: Metall. Trans. 3 (1972) 709

298. Karim, A.: Ultrafine Grain Metals. New York: Syracuse University Press 1970, p. 295

299. Lee, D.: Scripta Metall. 3 (1969) 893

300. Mukherjee, A.K.: Mater. Sci. Eng. 8 (1971) 83

301. Gandhi, C.; Rama Rao, P.; Taplin, D.M.R.: Met. Sci. 12 (1978) 30

302. Aubakirov, G.O.; Presnyakov, A.A.; Chernousova, K.T.: Russ. Metallurgy (4) (1969) 157

303. Melton, K.N.; Cutler, C.P.; Edington, J.W.: Scripta Metall. 9 (1975) 515

304. Homer, C.; Lechten, J.P.; Baudelet, B.: Metall. Trans. 8A (1977) 1191

305. Griffiths, P.; Hammond, C.: Scripta Metall. 7 (1973) 793

306. Griffiths, P.; Hammond, C.: Titanium Science and Technology Vol.2. New York: Plenum Press 1973, p. 1155

307. Padmanabhan, K.A.: Scripta Metall. 7 (1973) 787

308. Matsuki, K.; Morita, H.; Yamada, M.; Murakami, Y.: Met. Sci. 11 (1977) 156

309. Watts, B.M.; Stowell, M.J.; Baikie, B.L.; Owen, D.G.E.: Met. Sci. 10 (1976) 198

310. Kaibyshev, O.A.; Kazachkov, I.V.; Rozenberg, V.M.: Phys. Met. Metallog. 36(6) (1973) 103

311. Jones, R.B.; Johnson, R.H.: Trans. Am. Soc. Met. 59 (1966) 356; see also reply by Avery, D.H.; Backofen, W.A.: Trans. Am. Soc. Met. 59 (1966) 359

312. Benedek, R.A.; Doherty, R.H.: Scripta Metall. 8 (1974) 675

313. Langdon, T.G.: Scripta Metall. 4 (1970) 693

314. Mohamed, F.A.; Langdon, T.G.: Phil. Mag. 32 (1975) 697

315. Aldrich, J.W.; Avery, D.H.: Ultrafine Grain Metals. New York: Syracuse Univ. Press 1970, p. 397

316. Packer, C.M.; Sherby, O.D.: Trans. Am. Soc. Met. 60 (1967) 21

317. Dawson, T.H.: Metall. Trans. 3 (1972) 3201

318. Gifkins, R.C.; Snowden, K.U.: Trans. Met. Soc. AIME 239 (1967) 910

319. Padmanabhan, G.; Padmanabhan, K.A.: Phil. Mag. 33 (1976) 371

320. Smithells, C.J.: Metals Reference Book Vol. II. London: Butterworths 1967, pp. 644, 645, 682

321. Turton, D.; Nicholson, R.B.: Grain Boundary Sliding in High Temperature Deformation. London: Institute of Physics and Institute of Metals Discussion Meeting April 1971

322. Alden, T.H.: J. Aust. Inst. Met. 14 (1969) 207

323. Maulik, P.; Padmanabhan, K.A.: J. Mater. Sci. 10 (1975) 1646

324. Cook, R.C.: M.A.Sc. Thesis, Univ. British Columbia 1968

325. Francois, D.; Sherby, O.D.: J. Met. 20 (1968) 106A

326. Johnson, R.H.: Metall. Reviews 15 (1970) 115

327. Reichman, S.H.; Castledine, B.W.; Smythe, J.W.: SAE Report No.700140. Automotive Engineering Cong. Jan. 1970

328. Libanati, C.M.; Dyment, F.: Acta Metall. 11 (1963) 1263

329. Murdock, J.F.; Lundy, T.S.; Stansbury, E.E.: Acta Metall. 12 (1964) 1033

330. Murarka, S.P.; Agarwala, R.P.: Acta Metall. 12 (1964) 1096

331. Pahutová, M.; Hostinský, T.; Čadek, J.: Ryš, P.: Phil. Mag. 20 (1969) 975

332. Zhurkov, S.N.: Proc. Int. Conf. Fracture. Sendai Japan: 1965, p. 311

333. Watts, B.M.; Stowell, M.J.: Personal Communication

334. Padmanabhan, K.A.: Phys. Stat. Solidi (a) 22 (1974) K95

335. Kaibyshev, O.A.; Kazachkov, I.V.; Salikov, S.Ya.: Acta Metall. 26 (1978) 1887

336. Takeoka, K.; Oelschlaegel, D.; Izumi, O.: J. Jap. Inst. Met. 35(8) (1971) 78

337. Otsuka, M.; Miura, Y.; Horiuchi, R.: Scripta Metall. 8 (1974) 1405

338. Taplin, D.M.R.; Chandra, T.: J. Mater. Sci. 10 (1975) 1642

339. Morrison, W.B: Trans. Am. Soc. Met. 60 (1967) 352

340. Wray, P.J.: J. Appl. Phys. 40 (1969) 4018

341. Presnyakov, A.A.; Starikova, G.V.: Phys. Met. Metallog. 12 (6) (1961) 84

342. Headley, T.J.; Kalish, D.; Underwood, E.E.: Ultrafine Grain Metals. New York: Syracuse Univ. Press 1970, p. 325

343. Novikov, I.I.; Portnoy, V.K.; Terentieva, T.E.: Acta Metall. 25 (1977) 1139

344. Matsuki, K.; Ueno, Y.; Yamada, M.: J. Jap. Inst. Met. 38 (1974) 219

345. Karim, A.; Holt, D.L.; Backofen, W.A.: Trans. Met. Soc. AIME 245 (1969) 2421

346. Lorimer, G.W.; Nasir, M.J.; Nicholson, R.B.; Nuttall, K.; Ward, D.E.; Webb, J.R.: Electron Microscopy and Structure of Materials (ed. Thomas, G.). Berkeley: Univ. California Press 1972, p. 222

347. Price, R.J.; Kelly, A.: Acta Metall. 12 (1964) 159

348. Jones, R.L.; Kelly, A.: Acta Metall. 15 (1967) 438

349. Kelly, A.; Nicholson, R.B.: Progr. Mater. Sci. 10 (1963) 208, 209, 296,331

350. Dingley, D.J.: Proc. III Annual SEM Symp. IIT Illinois 1970, p. 329

351. Holtz, W.; Ruedl, E.; Schiller, P.: J. Mater. Sci. 10 (1975) 2003

352. Kobayashi, Y.; Ishida, Y.; Kato, M.: Scripta Metall. 11 (1977) 51

353. Bricknell, R.H.; Edington, J.W.: Acta Metall. 25 (1977) 447

354. Lee, D.: Scripta Metall. 4 (1970) 341

355. Langdon, T.G.; Gifkins, R.C.: Scripta Metall. 4 (1970) 337

356. Dingley, D.J.; Pope, P.J.: Modern Metallography in Metallurgy. Cambridge July 1972

357. Yadava, R.K.; Padmanabhan, K.A.: Mater. Sci. Eng. 37 (1979) 127

358. Gifkins, R.C.: Met. Sci. J. 7 (1973) 15

359. Gifkins, R.C.; Langdon, T.G.: Scripta Metall. 4 (1970) 563

360. Aigeltinger, E.H.: J. Mater. Sci. 9 (1974) 644

361. Kaibyshev, O.A.; Kazachkov, I.V.: Phys. Met. Metallog. 32 (1) (1971) 222

362. Heubner, U.; Matucha,K.; Sandig, H.: Z. Metallk. 63 (1972) 607

363. Cagnon, M.; Suery, M.; Eberhardt, A.; Baudelet, B.: Acta Metall. 25 (1977) 71

364. Ahlquist, C.M.; Menezes, R.A.: Mater. Sci. Eng. 7 (1971) 223

365. Langdon, T.G.; Mohamed, F.A.: Fracture 1977 Vol.2 (ed. Taplin, D.M.R.). Ontario: Univ. Waterloo Press 1977, p. 525

366. Ishikawa, H.; Bhat, D.G.; Mohamed, F.A.; Langdon, T.G.: Metall. Trans. 8A (1977) 523

367. Ridley, N.; Livesey, D.W.: Fracture 1977 Vol.2 (ed. Taplin, D.M.R.). Univ. Waterloo Press 1977, p. 533

368. Livesey, D.W.; Ridley, N.: Metall. Trans. 9A (1978) 519

369. Humphries, C.W.; Ridley, N.: J. Mater. Sci. 12 (1977) 851

370. Taplin, D.M.R.; Smith, R.F.: Fracture 1977 Vol. 2 (ed. Taplin, D.M.R.). Ontario: Univ. Waterloo Press 1977, p. 541

371. Shei, S.; Langdon, T.G.: J. Mater. Sci. 13 (1978) 1084

372. Woodthorpe, J.; Pearce, R.: Met. Sci. 11 (1977) 103

373. Thomason, P.F.: Met. Sci. J. 3 (1969) 139

374. Christian, J.W.: The Theory of Transformation in Metals and Alloys. Oxford; Pergamon 1965, p. 489.

375. Gifkins, R.C.: Scripta Metall. 7 (1973) 27

376. Hart, E.W.: Acta Metall. 15 (1967) 1545

377. Hazzledine, P.M.; Newbury, D.E.: 3rd I.C.S.M.A. London: Institute of Metals 1973, p. 202

378. Mohamed, F.A.; Langdon, T.G.: Phys. Stat. Solidi(a) 33 (1976) 375

379. Shei,S.; Langdon, T.G.; Acta Metall. 26 (1978) 639

380. Shei, S.; Langdon, T.G.: Acta Metall. 26 (1978) 1153

381. Padmanabhan, K.A.: Mater. Sci. Eng. 40 (1979) 285

382. Nabarro, F.R.N.: Report on Conference on Strength of Solids. London: Physical Society 1948, p. 75

383. Herring, C.: J. Appl. Phys. 21 (1950) 437

384. Coble, R.L.: J. Appl. Phys. 34 (1963) 1679

384. Coble, R.L.: J. Appl. Phys. 34 (1963) 1679

385. Chaudhari, P.: Metall. Trans. 5 (1974) 1692

386. Beere, W.: J. Mater. Sci. 12 (1977) 2093

387. Beere, W.: Phil. Trans. Roy. Soc. A288 (1978) 177

388. Gifkins, R.C.: Metall. Trans. 7A (1976) 1225

389. Gittus, J.H.: J. Eng. Mater. Technol. 99 (1977) 244

390. Gittus, J.H.: Phil. Trans. Roy. Soc. A288 (19/8) 121

391. Osipov, K.A. : Aspects of the Theory of Heat Resistance of Metals and Alloys. Moscow : Akad. Nauk SSSR 1960 (English translation Jerusalem : Israel Programme for Scientific Translations 1961)

392. Gurov, K.P.; Shorshorov, M.Kh.; Tikhonov, A.S.; Nadirashvili,N.I.: Phys.Met.Metallog. 34 (1972) 1145

393. Presnyakov,A.A.; Azhibayeva,A.S.M.; Soymin,N.Ya.; Aubakirova,R.K.; Karpenyuk,A.N.: Fiz.Met.Metalloved. 32 (1971) 654

394. Kroener,E. : Arch.Rational Mech.Anal. 4 (1960) 273

395. Hay,K.A.; Pascoe,R.T. : J.Mater.Sci.9 (1974) 1285

396. Stevens,R.N. : Phil.Mag. 23 (1971) 265

397. Cannon,W.R. : Phil.Mag. 25 (1972) 1489

398. Debye,P. : J.Chem.Phys. 14 (1946) 636

399. Greenwood,G.W. : Scripta Metall. 4 (1970) 171

400. Gifkins,R.C. : Scripta Metall. 10 (1976) 433

401. Tietz,T.E.; Dorn,J.E. : Trans.Met.Soc.AIME 206 (1956) 156

402. Huang,H.I.; Sherby,O.D.; Dorn,J.E. : Trans.Met.Soc.AIME 206 (1956) 1385

403. Mukherjee,A.K.; Bird,J.E.; Dorn,J.E. : Trans.Am.Soc.Met. 62 (1969) 155

404. Blum,N.; Reppich,B. : Acta Metall. 17 (1969) 959

405. Garofalo,F. : Fundamentals of Creep and Creep Rupture in Metals. New York : Macmillan 1965, pp.127-199

406. Padmanabhan,K.A. : Scripta Metall. 7 (1973) 137

407. Avery,D.H.; Backofen,W.A. : Trans.Am.Soc.Met. 59 (1966) 359

408. Nichols,F.A. : Mater.Sci.Eng. 8 (1971) 108

409. Yavari,P.; Langdon,T.G. : Scripta Metall. 11 (1977) 863

410 Gibbs, G.B.: Mem.Sci.Rev.Met. 62 (1965) 781

411. Lifshitz,I.M. : Sov.Phys.JETP 17 (1963) 909

412. Ashby,M.F.; Raj,R.; Gifkins,R.C. : Scripta Metall. 4 (1970) 737

413. Raj,R.; Ashby,M.F. : Metall. Trans. 2 (1971) 1113

414. Ashby,M.F. : Scripta Metall. 3 (1969) 837

415. Smith,C.I.; Norgate,B.; Ridley,N. : Scripta Metall. 8 (1974) 159

416. Spingarn,J.R.; Nix, W.D. : Acta Metall. 26 (1978) 1389

417. Ashby,M.F.; Edward,G.H.; Davenport,J.; Verrall,R.A.: Acta Metall. 26 (1978) 1379

418. Sherby,O.D.: Science Journal 5(6) (1969) 75

419. McLean,D.: Trans.Met.Soc. AIME 242 (1968) 1193

420. Kaibyshev,O.A.; Kazachkov,I.V.: Phys.Met.Metallog. 32 (1971) 222

421. Baero,G. : Z. Metallk. 63 (1972) 384

422. Schmidt-Whitley,R.D. : Scripta Metall. 8 (1974) 49

423. Schmidt-Whitley,R.D.: Scripta Metall. 8 (1974) 927

424. Young,C.M.; White,K.A.; Sherby,O.D.: Scripta Metall. 8 (1974) 923

425. Krausz,A.S. : Acta Metall. 16 (1968) 897

426. Krausz,A.S. : Scripta Metall. 5 (1971) 783

427. Krausz,A.S. : J.Met. 25 (4) (1973) 24

428. Krausz,A.S. : Int.J.Fracture 10 (1974) 503

429. Stowell,M.J. : Personal Communication

430. Mott,N.F. : Phil.Mag. 44 (1953) 742

431. Mott,N.F. : Proc.Phys.Soc. B64 (1951) 729

432. Friedel,J. : Dislocations, London : Pergamon 1964, pp.82,279,315

433. Buchanan,E.R.: Trans.Am.Soc.Met. 61 (1968) 852

434. Alden,T.H. : Acta Metall. 17 (1969) 1435

435. Burton,B.: Scripta Metall. 3 (1969) 247

436. Dunlop,G.L.; Taplin,D.M.R. : Scripta Metall. 3 (1969) 641

437. Aldrich,J.W.; Avery,D.H. : II Int. Conf. on Strength of Metals and Alloys. Metals Park Ohio : Am.Soc.Metals Vol.3 1970, p.1094

438. Beere,W. : Scripta Metall. 12 (1978) 337

439. Gifkins,R.C.; Langdon,T.G. : Mater.Sci.Eng. 36 (1978) 27

440. Horton,C.A.P.; Beevers,C.J. : Acta Metall. 16 (1968) 733

441. Roy,B.; Ranganathan,S. : Trans.Ind.Inst.Met. 29 (1976) 62

442. Mott,N.F. : Proc.Phys.Soc. 60 (1948) 391

443. Brandon,D.G.; Ralph,B.; Ranganathan,S.; Wald,M.S. : Acta Metall. 12 (1964) 813

444. Bollmann,W. : Crystal Defects and Crystalline Interfaces. Berlin : Springer-Verlag 1970, pp.148-154

445. Hirth,J.P.; Baluffi,R.W. : Acta Metall. 21 (1973) 929

446. Lin,T.L.; McLean,D. : Met.Sci.J. 2 (1968) 108

447. Gleiter,H.; Hornbogen,E.; Baero,G.: Acta Metall. 16 (1968) 1053

448. Ashby,M.F. : Surface Science 31 (1972) 498

449. Gates,R.S. : Acta Metall. 21 (1973) 855

450. Ishida,Y.; Brown,M.H. : Acta Metall. 15 (1967) 857

451. Kegg,G.R.; Horton,C.A.P.; Silcock,J.M. : Phil.Mag. 27 (1973) 1041

452. Gleiter,H. : Acta Metall. 17 (1969) 853

453. Weertman,J. : J.Appl.Phys. 26 (1955) 1213

454. Crussard,C.; Friedel,J. : Creep and Fracture of Metals at Elevated Temperatures. London: H.M.S.O. 1956,p.243

455. Horton,C.A.P.; Thompson,N.B.W.; Beevers,C.J. : Met.Sci.J. 2 (1968) 19

456. Halliday, M.D.; Horton,C.A.P.; Beevers,C.J. : Met.Sci.J. 3 (1969) 195

457. Langdon,T.G. : Phil.Mag. 22 (1970) 689

458. Taplin,D.M.R. : Met.Eng. Quart. 10 (1970) 31

459. Langdon,T.G. : Mater.Sci.Eng.7 (1971) 117

460. McLean,D. : Met.Sci.J. 6 (1972) 109

461. Gleiter,H.; Baero,G. : Mater.Sci.Eng. 2 (1967) 224

462. Baero,G.; Gleiter,H.; Hornbogen,E.: Mater.Sci.Eng. 3 (1968-69) 92

463. Buzzichelli,G.; Mascanzoni,A. : Phil.Mag. 24 (1971) 497

464. McLean,D. : Grain Boundary Structure. London : The Institute of Metallurgists Review Course Ser.2(7) 1971, pp.1-7,44

465. Pumphrey,P.H.; Gleiter,H. : Phil.Mag. 36 (1977) 1099

466. Pumphrey,P.H. : Grain Boundary Structure and Properties (eds. Chadwick,G.A.; Smith,D.A.). London : Academic Press 1976, p.139

467. Pumphrey,P.H.; Gleiter,H. : Phil.Mag. 30 (1974) 593

468. Pumphrey, P.H.; Gleiter,H.: Phil.Mag. 32 (1975) 881

469. Misra,S.; Suryanarayana,C.; Ranganathan, S. : Ind.J.Technol. 11 (1973) 435

470. Nature 243 (1973) 222

471. Gleiter,H. : Phil.Mag. 36 (1977) 1109

472. Aaron,H.B.; Bolling,G.F. : Grain Boundary Structure and Properties (eds. Chadwick,G.A.; Smith,D.A.). London: Academic Press 1976, p.107

473. Pond,R.C.; Smith,D.A. : Phil.Mag.36 (1977) 353

474. Ashby, M.F.; Spaepen, F.; Williams, S. : Acta Metall. 26 (1978) 1647

475. Pond,R.C.; Smith, D.A.; Vitek, V. : Scripta Metall. 12 (1978) 699

476. Bernall, J.D. : Proc.Roy.Soc. A280 (1964) 299

477. Mott, N.F.; Davis, E.A. : Electronic Processes in Non-Crystalline Materials. Oxford : Clarendon 1971, p. 77

478. Kê, T.S. : J.Appl.Phys. 20 (1949) 274

479. Harris, J.E.: Grain Boundary structure. London : The Institution of Metallurgists Review Course Ser.2(7) 1971,p.18

480. Walter,J.L.; Cline, H.E. : Trans.Met.Soc. AIME 242 (1968) 1823

481. Gjostein, N.A. : Diffusion. Metals Park Ohio : Am.Soc. Metals 1973, p.272

482. Lange,W.; Bergner,D. : Phys.Stat. Solidi 2 (1962) 1410

483. Orowan, E. : Z. Phys. 98 (1936) 386

484. Padmanabhan, K.A.; Davies, G.J. : 7th Proc.Int.Congr. Rheology. Gothenburg : Swedish Society of Rheology 1976, p.326

485. Weertman, J.; Weertman, J.R. :Physical Metallurgy (ed.Cahn,R.W.). Amsterdam : North Holland Publishing Co. 1965, p.793

486. Mohamed, F.A.; Langdon, T.G. : Scripta Metall. 10 (1976) 759

487. Ashby, M.F. : Acta Metall. 20 (1972) 887

488. Langdon, T.G. ; Mohamed, F.A. : Mater.Sci.Eng. 32 (1978) 103

489. Fleck, R.G.; Cocks, G.J.; Taplin, D.M.R. : Metall.Trans. 1 (1970) 3415

490. Holt, D.L.; Backofen, W.A. : Trans.Am.Soc.Met. 60 (1967) 275

491. Alden, T.H. : Trans.Am.Soc.Met. 60 (1967) 274

492. Kaibyshev, O.A.; Valiev, R.Z.; Astanin, V.V. : Phys.Stat. Solidi (a) 35 (1976) 403

493. Packer, C.M. : Ph.D. Thesis, Stanford Univ. 1967

494. Clinard, F.W.; Sherby, O.D. : Acta Metall. 12 (1964)911

495. Lehr, P. : Compt.Rend. 242 (1956) 1172

496. de Jong, M.; Rathenau, G.W. : Acta Metall. 9 (1961) 889

497. Russell, A.M.; Abbott, R.C. : J.Appl.Phys. 29 (1958) 1130

498. Kot, R.A.; Weiss, V. : Trans.Am.Soc.Met. 60 (1967) 566

499. Hoke, J.H.; Leitch, R.H.; Matthews, F.D. : Trans.Am.Soc.Met. 60 (1967) 566

500. Petsche, S.; Stangler, F. : Z.Metallk. 62 (1971) 601

501. Buckley,S.N.; Harding, A.G.; Waldron, M.B. : J.Inst.Met. 87 (1958-59) 150

502. Stobo, J.J. : J.Nucl.Mater. 2(2) (1960) 97

503. Petsche, S.; Stangler, F. : Z.Metallk. 62 (1971) 606

504. Bush, R.H.; Bokros, I.C. : Acta Metall. 12 (1964) 102

505. Arbuzova, I.A.; Khandros, L.G. : Phys.Met.Metallog. 17(3) (1964) 68

506. Ankara, O.A.; Sastri, A.S.; West, D.R.F. : JISI 204 (1966) 509

507. Logunov, V.Ya.; Filippov, M.A. : Phys.Met.Metallog. 38(3) (1974) 164

508. Khachin, V.N.; Gyunter, V.E.; Paskal, Yu.I. : Phys.Met.Metallog. 44(4) (1978) 183

509. Sattler, H.P.; Wassermann, G. : J.Less-Common Met. 28 (1972) 119

510. Porter, L.F.; Rosenthal, P.C. : Acta Metall. 7 (1959) 504

511. Ankara, O.A.; West, D.R.F. : JISI 205 (1967) 36

512. Guy, A.G.; Pavlick, J.E. : Trans.Met.Soc.AIME 221 (1961) 802

513. Buehler, W.J.; Wiley, R.C. : Trans.Am.Soc.Met. 55 (1962) 269

514. Bressanelli, J.P.; Moskowitz, A. : Trans.Am.Soc.Met. 59 (1966) 223

515. Otsuka, K.; Wayman, C.M.; Nakai, K.; Sakamoto, H.; Schimizu, K. : Acta Metall. 24 (1976) 207

516. Chaklader, A.C.D.; Baker, V.T. : Amer.Ceram.Soc.Bul. 44 (1965) 258

517. Newey, C.W.A.; Radford, K.C. : Anisotropy in single Crystal Refractory Compounds (eds. Vahldick, F.W.; Mersol, S.M.). London : Plenum Press 1968, p.321

518. Steiner, C.J.P.; Spriggs, R.P.; Hasselman, D.P.H.: J.Amer.Ceram.Soc. 55(2) (1972) 115

519. Chaklader, A.C.D. : Nature 197 (1963) 791

520. Lainer, D.I.; Tsypin, M.I.; Novikov, A.V.; Shevakin, Yu.F.; Sollertinskaya, E.S.; Afonin, M.P. : Sov.Phys.Dokl. 18 (3) (1973) 195

521. Chiswik, H.H. : Trans.Am.Soc.Met. 49 (1957) 622

522. Zegler, S.T.; Mayfield, R.M.; Mueller, M.H. : Trans.Am.Soc.Met. 50 (1958) 905

523. Mayfield, R.M. : Trans.Am.Soc.Met. 50 (1958) 926

524. Lobb, R.C.; Sykes, E.C.; Johnson, R.H. : J.Inst.Met. 96 (1968) 262

525. Lobb, R.C.; Sykes, E.C.; Johnson, R.H. : Met.Sci.J. 6 (1972) 33

526. Cottrell, A.H. : Metall.Reviews 1 (1956) 479

527. Roberts, A.C.; Cottrell, A.H. : Phil.Mag. 1 (1956) 711

528. Angerman, C.L.; Caskey Jr., G.R. : J.Nucl. Mater. 13(2) (1964) 182

529. Hudson, B.; Westmacott, K.H.; Makin, M.J. : Phil.Mag. 7 (1962) 377

530. Hudson, B. : Phil.Mag. 10 (1964) 947

531. Harrison, J.W. : J.Nucl.Mater. 23 (1967) 139

532. Gurov, K.P.; Shorshorov, M.Kh.; Tikhonov, T.S.; Nadirashvili, N.I.: Fiz.Met.Metalloved. 34 (1972) 1148

533. Tikhonov,A.S.; Gurov, K.P.; Shorshorov, M.Kh.; Prokhorova, I.I. : Phys.Met.Metallog. 36(2) (1973) 7

534. Sherby, O.D. : Acta Metall. 10 (1962) 135

535. Burke, J.E.; Turkalo, A.M. : Trans.Am.Soc.Met. 50 (1958) 943

536. Lozinsky, M.G. : Acta Metall. 9 (1961) 689

537. Sherby, O.D.; Goldberg, A. : Acta Metall. 9 (1961) 510

538. Sherby, O.D.; Lytton, J.L. : Trans.Met.Soc.AIME 206 (1956) 928

539. Clinard, F.W.; Sherby, O.D. : Trans.Met.Soc. AIME 233 (1965) 1975

540. Grupen, W.B.; Sines, G.: Acta Metall. 13 (1965) 527

541. Froes, F.H.; Honeycombe, R.W.K.; Warrington, D.H. : Acta Metall. 15 (1967) 157

542. Froes, F.H.; Warrington,D.H. : Trans.Met.Soc. AIME 245 (1969) 2009

543. Gilman, J.J. : J.Appl.Phys. 36 (1965) 3195

544. Thomsen, T.H.; Holt, D.L.; Backofen, W.A. : Met.Eng.Quart. 10(2) (1970) 1

545. Hestbech, J.; Langer, E.W.; Rosen, A. : J.Inst.Met. 99 (1971) 306

546. Belk, J.A. : Int.J.Mech.Sci. 17 (1975) 505

547. Tang, S.; Robbins, T.L. : J.Eng.Mater.Technol. (Trans. ASME) 96A (1974) 77

548. Cornfield, G.C.; Johnson, R.H. : Int.J.Mech.Sci. 12 (1970) 479

549. Cocks, G.J.; Rowbottom, G.; Taplin, D.M.R. : Met.Technol. 3 (1976) 332

550. Johnson, W.; Al-Naib, T.Y.M.; Duncan, J.L. : J.Inst.Met. 100 (1972) 45

551. Clemas, G.G.W.; Al-Hassani, S.T.S.; Johnson, W. : Int.J. Mech.Sci. 17 (1975) 711

552. Duncan, J.L.; Johnson, W. : Int.J.Mech.Sci. 10 (1968) 157

553. Fields Jr., D.S. : IBM J.Res.Dev. 9 (1965) 134

554. Davis,D.W.; Osborne, C.F.; Brooks, G.J. : Met.Eng.Quart. 13(4) (1973) 5

555. Turner, L.W. : Proc.Conf.Technology of Engineering Manufacture (I.Mech.E.) London 1958, p.198

556. Mortimer, J. : The Engineer 232 (1971) 34

557. North, D. : Sheet Metal Industries 47 (1970) 13

558. Mamalis, A.J.; Johnson, W.; Lewis, J. : Met.Technol. 4 (1977) 160

559. Petersen, W.A.; Lang, F.H. : Met.Eng.Quart. 15(1) (1975) 31

560. Oshita, T.; Takei, H. : J.Jap.Inst.Met. 36 (1972) 1081

561. Hawkins, R.; Belk, J.A. : Met.Technol. 3 (1976) 516

562. Oshita, T.; Takei, H. : J.Jap.Inst.Met. 38 (1974) 165

563. Saller, R.A.; Duncan, J.L. : J.Inst.Met. 99 (1971) 173

564. Stewart, M.J. : Can.Met.Quart. 12 (1973) 159

565. Balliett, R.W.; Forster, J.A.; Duncan, J.L. : Metall. Trans. 7A (1978) 1259

566. Frost, L.E. : Metal Progr. 97 (1970) 86

567. Moles, M.D.C.; Forster, J.A.; Duncan, J.L. : Can.Met.Quart. 15 (1976) 249

568. Hubert, J.F.; Kay, R.C. : Met.Eng.Quart. 13(4) (1973)1

569. Jovane, F.; Shabaik, A. ; Thomsen, E.G. : J.Eng.Ind. (Trans.ASME) 91 (1969) 657

570. Thomsen, E.G.; Lapsley Jr., J.T. : Proc.Soc.Exp.Stress Analysis 11 (1954) 59

571. Shabaik, A.; Kobayashi, S. : J.Eng.Ind. (Trans.ASME) 89 (1967) 339

572. Tang, S. : J.Basic Eng. (Trans.ASME) 93 (1971) 657

573. Oshita, T.; Takei, H. : J.Jap.Inst.Met. 35 (1971) 1120

574. Johnson, R.H. : Design Engineering (March 1969) 33

575. Athey, R.L.; Moore, J.B. : Powder Metallurgy for High Performance Applications (eds. Burke, J.J.; Weiss, V.). New York : Syracuse Univ. Press 1972, p.281

576. Gessinger, G.H.; Bomford, M.J. : Int.Metall.Reviews 19 (1974) 51

577. Holt, D.L. : Int.J.Mech.Sci. 12 (1970) 491; see also comment by Cornfield, G. : Int.J.Mech.Sci. 13(1971)81

578. Woo, D.M. : Int.J.Mech.Sci. 6 (1964) 303

579. Timoshenko, S.P. : The Theory of Plates and Shells. New York : McGraw Hill 1959, p.547

580. North, D. (comment on ref. 577) : Int.J.Mech.Sci. 13 (1971) 79

581. Alto, A.; Giorleo, G. : J.Eng.Ind. (Trans. ASME) 97 (1975) 1131

582. Padmanabhan, K.A.; Davies, G.J. : J.Mech.Phys.Sol. 18 (1970) 261.

583. Hundy, B.B. : Plasticity and Superplasticity. London : Institution of Metallurgists Review Course Ser. 2(3) 1969, p. 73

584. Dollar, D.L.; Clum, J.A.; Miller, R.E. : Corrosion 28 (1972) 296

585. Frommeyer, G. : Z.Metallk. 67 (1976) 361

586. Metallurgia 83 (1971) 55

587. Peters, D. : The Engineer 232 (1971) 43

588. Sumner, J. : The Engineer 237 (1973) 71

589. Sawle, R. : The Chartered Mechanical Engineer 25(12) (1978) 73

590. Hamilton, C.H.; Stacher, G.W. : Metal Progr. 109(3) (1976) 34

591. Sumner, J. : The Engineer 242 (1976) 11

592. Young, R.J.; Ziats, A.W.; Mehl, D.L. : Met.Eng.Quart. 13(4) (1973) 16

Subject Index

Author Index

Materials Research and Engineering

Editor: B. Ilschner

Volume 1

D. G. Altenpohl

Materials in World Perspective

Assessment of Resources, Technologies and Trends for Key Materials Industries

In collaboration with T. S. Daugherty
With contributions by M. Bever, J. C. Clark, H. Eldag, G. Friese, P. Kelterborn, I. Reznik, D. Spreng, F. R. Tuler
1980. 33 figures. XVI, 208 pages
ISBN 3-540-10037-7

From the contents: Role of Materials in the World Economy. – Present Structure and Future Trends in Key Materials Industries (Introduction, Iron and Steel, Aluminium, Copper, Cement and Concrete, Plastics, Wood and Wood Products, Advanced Materials). – Technology Planning as Part of Industry's Planning Process. – Key Issues for Technology Planning and Assessment. – Research and Development Opportunities. – Outlook. – Index.

This book deals with the entire materials cycle – from extraction or harvesting to processing, manufacture, use, and reuse or disposal. It covers the present status and ongoing developments in six key materials industries in both industrialized and developing countries.

Techno-economic trends, which are recognizable today, as well as important changes taking place from the mine through the refining stage an to finished products are outlined. The "problem triangle" of the materials industry – basic or raw materials, ecology, and energy – is discussed. Of specific importance are the impacts which a given material or technology can have the environment. Methods of assessing these impacts, which should be integrated into overall technology planning by the materials industry, are described.

This book discusses resources, industry's social responsibilities and limits-to-growth. An explanation is given for apposing views on constraints and growth, not only for the materials industry, but also for the automotive and packaging industries.

Thus, this book spotlights the interactions between different fields of technology and their interrelationship with different regions on earth.

Springer-Verlag
Berlin
Heidelberg
New York

Reine und angewandte Metallkunde in Einzeldarstellungen

Herausgeber: W. Köster

The series will be continued as "Materials Research and Engineering"

A Selection

19. Band
D. Altenpohl
Aluminium und Aluminiumlegierungen
1965. 662 Abbildungen.
XX, 899 Seiten
ISBN 3-540-03388-2

20. Band
H. Schreiner
Pulvermetallurgie elektrischer Kontakte
1964. 197 Abbildungen. VIII, 239 Seiten
ISBN 3-540-03200-2

21. Band
U. Zwicker
Titan und Titanlegierungen
Unter Mitarbeit zahlreicher Fachleute
1974. 387 Abbildungen. XI, 717 Seiten
ISBN 3-540-05233-X

22. Band
R. E. Hummel
Optische Eigenschaften von Metallen und Legierungen
Mit einer Einführung in die Elektronentheorie der Metalle
1971. 154 Abbildungen. VIII, 222 Seiten
ISBN 3-540-05234-8

23. Band
B. Ilschner
Hochtemperatur – Plastizität
Warmfestigkeit und Warmverformbarkeit metallischer und nichtmetallischer Werkstoffe
1973. 123 Abbildungen. IX, 314 Seiten
ISBN 3-540-05966-0

24. Band
Order-Disorder Transformations in Alloys
Proceedings of the International Symposium on Order-Disorder Transformations in Alloys
held September 3–6, 1973, Tübingen
Sponsor: Deutsche Gesellschaft für Metallkunde
Editor: H. Warlimont
1974. 304 figures. VIII, 554 pages
ISBN 3-540-06766-3

25. Band
W. Kurz, P. R. Sahm
Gerichtet erstarrte eutektische Werkstoffe
Herstellung, Eigenschaften und Anwendungen von In-situ-Verbundwerkstoffen
1975. 203 Abbildungen. VII, 345 Seiten
ISBN 3-540-06998-4

26. Band
Gase und Kohlenstoff in Metallen
Herausgeber: E. Fromm, E. Gebhardt
1976. 510 Abbildungen.
XVIII, 747 Seiten
ISBN 3-540-07255-1

Springer-Verlag
Berlin Heidelberg New York

GPSR Compliance

The European Union's (EU) General Product Safety Regulation (GPSR) is a set of rules that requires consumer products to be safe and our obligations to ensure this.

If you have any concerns about our products, you can contact us on ProductSafety@springernature.com

In case Publisher is established outside the EU, the EU authorized representative is:

Springer Nature Customer Service Center GmbH
Europaplatz 3
69115 Heidelberg, Germany

Batch number: 10640288

Printed by Printforce, the Netherlands